Springer Series on
ATOMIC, OPTICAL, AND PLASMA PHYSICS 41

Springer Series on
ATOMIC, OPTICAL, AND PLASMA PHYSICS

The Springer Series on Atomic, Optical, and Plasma Physics covers in a comprehensive manner theory and experiment in the entire field of atoms and molecules and their interaction with electromagnetic radiation. Books in the series provide a rich source of new ideas and techniques with wide applications in fields such as chemistry, materials science, astrophysics, surface science, plasma technology, advanced optics, aeronomy, and engineering. Laser physics is a particular connecting theme that has provided much of the continuing impetus for new developments in the field. The purpose of the series is to cover the gap between standard undergraduate textbooks and the research literature with emphasis on the fundamental ideas, methods, techniques, and results in the field.

Vols. 10–35 of the former Springer Series on Atoms and Plasmas are listed at the end of the book

A.Y. Varaksin

Turbulent Particle-Laden Gas Flows

With 105 Figures

 Springer

Professor Dr. Aleksej Y. Varaksin
Institute for High Temperatures
Russian Academy of Sciences
13/19 Izhorskaya street, 125412 Moscow, Russia
E-mail: varaksin_a@mail.ru

ISSN 1615-5653

ISBN-10 3-540-68053-5 Springer Berlin Heidelberg New York

ISBN-13 978-3-540-68053-6 Springer Berlin Heidelberg New York

Library of Congress Control Number: 2006940763

Springer is a part of Springer Science+Business Media.

springer.com

© Springer-Verlag Berlin Heidelberg 2007

Typesetting and production: SPi Publisher Services
Cover design: eStudio Calmar Steinen

Printed on acid-free paper   SPIN: 11930761   57/3100/SPI - 5 4 3 2 1 0

Preface

Turbulent air flows, which carry solid particles, occur widely in nature and find application in numerous fields of human activities. For several decades now, gas–solid heterogeneous flows have been attracting researchers' attention. Quite extensive theoretical and experimental data have been accumulated to date, which are devoted to diverse aspects of gas dynamics and thermophysics of such flows [1–34].

The presence of even an insignificant amount of a disperse impurity in flows of gas media may cause undesirable effects. As a result, the study of such flows and the development of mechanics of heterogeneous media become extremely urgent.

In spite of great interest shown by numerous teams of researchers the world over in studying heterogeneous flows and of the large number of papers on the subject, the currently available theory of multiphase turbulent flows is inadequate. This is apparently due to two reasons. First, the theory of single-phase turbulent flows of continuous media is at present far from being complete. Second, the addition of a disperse impurity in the form of particles to a turbulent flow (complex as this flow is) causes a serious complication of the flow pattern. This is first of all associated with the great diversity of the properties of particles being introduced, which results in the realization of numerous flow modes of the gas suspension. By varying the concentration of particles (which is the main extensive characteristic of heterogeneous flows), one can change both qualitatively the parameters of initial flow and of particle motion and accomplish qualitative restructuring of the flow (for example, the transition of laminar flow mode to turbulent, as well as inverse effect, i.e., relaminarization of flow). Because of this, the experimental and theoretical investigation techniques employed in the classical mechanics of single-phase continuous media more often than not are unfit for use in studying heterogeneous flows. The available experimental data are often fragmentary and contradictory, while the physical concepts and developed mathematical models cannot be recognized as adequate. The foregoing factors impede the development of the mechanics of heterogeneous media. Nevertheless, the practical needs and the logic of scientific development demand continuous perfection of the theory of heterogeneous flows.

This monograph deals with problems associated with the hydrodynamics of turbulent flows of air in the presence of solid particles in pipes (channels) and under conditions of flow past bodies. The closest in content to this book is the monograph by Shraiber et al. [22]. The problems treated there have found further development in this book.

The first, introductory, chapter contains brief information about turbulent single-phase wall flows, which is essential for the understanding of the problems treated in the book. This information is borrowed from the available literature sources and is in no way original. Also given in the first chapter are the main characteristics of gas flows with particles, and the suggested classification of heterogeneous turbulent flows is described.

Two subsequent chapters deal with basic approaches and methods of mathematical and physical simulation of heterogeneous flows. The entire history of development of natural science confirms the mutual importance and interdependence of theoretical and experimental investigation techniques. In constructing the theory of any physical phenomenon (however, complex or simple it might seem to be at first glance), one must not underestimate the importance of some or other methods of investigation. The foregoing is well supported by the entire history of development of the theory of turbulent single-phase and multiphase flows. In recent years, in view of rapid development of computer equipment, mathematical simulation techniques (numerical methods) have come to play an important part in the development of the theory of two-phase flows. The use of these methods enables one to solve systems of complex differential equations and obtain detailed information about the fine structure of heterogeneous flows. Rapid progress in computer development gave a powerful impetus to the development of experimental investigation techniques. The use of high-speed processors makes possible the measurement of fine structure characteristics of heterogeneous flows in real time.

The second chapter contains the description of presently available methods of mathematical simulation of gas flows with solid particles. Analysis is made of the validity of some or other approaches for studying particleladen flows of different classes in accordance with the classification given in the first chapter.

In the third chapter, the problems of physical simulation of heterogeneous flows are treated. The fundamentals of laser Doppler anemometry (LDA) are described: during the last several decades, this method has become one of the most extensively used means of fine diagnostics of single-phase flows. A wide range of metrological problems, which arise during investigations of heterogeneous flows using this method are discussed. Such problems include the optimization of the parameters of the optoelectronic system of laser Doppler anemometers for measuring the instantaneous velocity of large particles of the dispersed phase, the development of the procedure for correct measurement of the velocity of substantially polydisperse particles, the elaboration of the principles of signal selection required for studying the inverse effect of particles on the characteristics of carrier air flow, the development of the procedure for measurement of the concentration of particles, and so on. Along with the description of the techniques employed in the LDA diagnostics of heterogeneous flows, much attention is given to the problems associated with the theoretical and experimental monitoring of the measurement results. Examples of experimental apparatuses for studying flows of air with particles are given at the end of the chapter, as well as the description of the principle of selection of characteristics of solid particles which are used in the investigation of heterogeneous flows as the disperse phase.

The fourth chapter treats the motion of disperse phase and characteristic features of interphase processes under conditions of gas flow with solid particles in channels (pipes). The results of experimental investigations of gas-solid flows in channels are described for heterogeneous flows of different classes. Analysis is made of the data of measurements of distributions of averaged and fluctuation velocities of particles in a wide range of the particle concentration. Special attention is given to the experimental and theoretical study of one of the fundamental problems in the mechanics of multiphase media, namely, the problem of modification by the particles of the turbulent energy of the carrier phase. Analysis is made of the results of experimental investigation involving, for the first time in the "pure" state (the presence of particles did not affect the profile of averaged velocity of the carrier phase), a study of the process of additional dissipation of turbulence in a flow with relatively low-inertia particles. The modification of the turbulent energy by particles is studied theoretically. The mathematical model, which enables one to determine the values of additional generation and dissipation of turbulence in flows with particles is described. The calculations involving the use of this model made possible the generalization of the available data on the modification of the turbulent energy of carrier gas by particles in a wide range of variation of the concentration and inertia of these particles.

In the fifth chapter, the characteristic features of gas flows with particles past bodies are described. Analysis is made of the available data on the behavior of particles in the vicinity of the critical point of bodies of different shapes subjected to flow, as well as on the effect of particles on the characteristics of the carrier phase. The effect of various factors (particle inertia, Saffman force, etc.) on the deposition of particles is treated. Much attention is given to the description of singular features of heterogeneous flow in the boundary layer developing along the surface of a body. The experimental data on the distribution of velocities of "pure" air, air with particles, and solid particles proper in all regions of the boundary layer developing along the surface of the model, i.e., laminar, transition, and turbulent regions, are treated and analyzed. It is demonstrated that the presence of particles in the flow precipitates the beginning of the laminar–turbulent transition. The effect of the particles on the intensity of turbulence of carrier air in the turbulent boundary layer is treated. The experimental data on the distribution of the velocities of incident particles and particles reflected from the body surface are described

VIII   Preface

and analyzed. The size of the region of existence of the "phase" of reflected particles is determined for the inertia of the dispersed phase varying in a wide range. The behavior of particles repeatedly interacting with the body surface is studied.

I am grateful to A.I. Leontiev, Member of the Russian Academy of Sciences, V.M. Batenin, Corresponding Member of the Russian Academy of Sciences, Yu.V. Polezhaev, Corresponding Member of the Russian Academy of Sciences, and Prof. Yu.A. Zeigarnik for their longstanding support and attention to this study, as well as to Profs. A.F. Polyakov and L.I. Zaichik for their participation in a number of investigations the results of which are used in this book. I am very thankful to H.A. Bronstein and S.G. Yankov for translation and preparation of manuscript.

Moscow, April 2007   *Aleksei Y. Varaksin*

Contents

Symbols

Dimensional quantities

D — pipe diameter, m

R — pipe radius, m

r — distance from the pipe axis, m

l — Prandtl–Nikuradse mixing length, m

ρ — density of gas, $\mathrm{kg\,m^{-3}}$

ρ_p — density of the particle material, $\mathrm{kg\,m^{-3}}$

d_p — particle diameter, m

x, y, z — axial, radial, and azimuthal Cartesian coordinates, m

x, r, φ — axial, radial, and azimuthal cylindrical coordinates, m

N — numerical concentration of particles, $\mathrm{m^{-3}}$

u_i — components of instantaneous velocity of gas, $\mathrm{m\,s^{-1}}$

v_i — components of instantaneous velocity of particle, $\mathrm{m\,s^{-1}}$

U_i — components of averaged velocity of gas, $\mathrm{m\,s^{-1}}$

V_i — components of averaged velocity of particle, $\mathrm{m\,s^{-1}}$

u_i' — components of fluctuation velocity of gas, $\mathrm{m\,s^{-1}}$

v_i' — components of fluctuation velocity of particle, $\mathrm{m\,s^{-1}}$

u_* — dynamic velocity, $\mathrm{m\,s^{-1}}$

t — instantaneous temperature of gas, K

t_p — instantaneous temperature of particle, K

T — averaged temperature of gas, K

T_p — averaged temperature of particle, K

t' — fluctuation temperature of gas, K

t_p' — fluctuation temperature of particle, K

p — instantaneous pressure, Pa

P — averaged pressure, Pa

p' — fluctuation pressure, Pa

μ — coefficient of dynamic viscosity, $\mathrm{N\,s\,m^{-2}}$

ν — coefficient of kinematic viscosity, $\mathrm{m^2\,s^{-1}}$

λ — thermal conductivity coefficient of gas, $\mathrm{W\,m^{-1}K^{-1}}$

λ_p – thermal conductivity coefficient of the particle material, $\text{W m}^{-1}\text{K}^{-1}$
C_p – heat capacity of gas, $\text{J kg}^{-1}\text{K}^{-1}$
C_{p_p} – heat capacity of the particle material, $\text{J kg}^{-1}\text{K}^{-1}$
a – thermal diffusivity of gas, m^2s^{-1}
g – acceleration of gravity, m s^{-2}
k – turbulent energy of gas, m^2s^{-2}
k_0 – turbulent energy of gas in the absence of particles, m^2s^{-2}
k_p – energy of velocity fluctuations of particles, m^2s^{-2}
ε – rate of dissipation of turbulent energy, m^2s^{-3}
τ – time, s
τ_p – time of dynamic relaxation of particle, s
τ_po – time of dynamic relaxation of Stokesian particle, s
τ_t – time of thermal relaxation of particle, s
τ_to – time of thermal relaxation of Stokesian particle, s
T_f – characteristic time of gas in averaged motion, s
T_L – characteristic time of gas in large-scale fluctuation motion, s
τ_K – characteristic time of gas in small-scale fluctuation motion
         (Kolmogorov timescale of turbulence), s
τ_w – shear stress on the wall, Pa

Dimensionless quantities

m – instantaneous mass concentration of particles
M – averaged mass concentration of particles
m' – fluctuation mass concentration of particles
φ – instantaneous volume concentration of particles
Φ – averaged volume concentration of particles
φ' – fluctuation volume concentration of particles
Re_D – Reynolds number
Re_x – local value of the Reynolds number, calculated
        by longitudinal coordinate
$\widetilde{Re}_\text{p}$ – instantaneous value of the Reynolds number of particle
Re_p – averaged value of the Reynolds number of particle
Re'_p – fluctuation value of the Reynolds number of particle
Stk_f – Stokes number in averaged motion
Stk_L – Stokes number in large-scale fluctuation motion
Stk_K – Stokes number in small-scale fluctuation motion
γ – Karman constant
C_D – coefficient of aerodynamic drag of particle
C_{x0} – body drag coefficient
C_{x_p} – coefficient of body drag due to the effect of particles
C_f – coefficient of friction

Indexes

$\langle \cdots \rangle$ – averaging over the cross-sectional area of the pipe (channel)

$(\overline{\cdots})$ – time averaging, relative value

$(\cdots)'$ – fluctuation value

Subscripts

c – value on the pipe (channel) axis

w – value on the pipe (channel) wall

0 – value in external flow, value in the absence of particles

m – modified value

1

Concise Information About Single-Phase and Heterogeneous Turbulent Flows

1.1 Preliminary Remarks

In subsequent chapters, data about the respective characteristics of flows in the absence of particles are used in describing the behavior of particles in turbulent flows and their effect on the characteristics of flow of continuous medium. Such data are given in this chapter. The data in Sects. 1.2 and 1.3 are borrowed from the monographs [2–4, 9, 12, 15, 17, 19, 20] which contain more detailed information on the problems being treated. Sections 1.4 and 1.5 are devoted to the description of the main characteristics of flows with solid particles and suggested classification of turbulent heterogeneous flows.

1.2 Equations of Single-Phase Turbulent Flows

Given in this section are basic equations which describe turbulent single-phase flows. Equations of continuity, motion, and energy for incompressible gas in terms of actual variables in the absence of external mass forces have the form

$$\sum_j \frac{\partial u_j}{\partial x_j} = 0, \tag{1.1}$$

$$\frac{\partial u_i}{\partial \tau} + \sum_j u_j \frac{\partial u_i}{\partial x_j} = -\frac{1}{\rho} \frac{\partial \rho}{\partial x_i} + \nu \sum_j \frac{\partial^2 u_i}{\partial x_j \partial x_j}, \tag{1.2}$$

$$\frac{\partial t}{\partial \tau} + \sum_j u_j \frac{\partial t}{\partial x_j} = a \sum_j \frac{\partial^2 t}{\partial x_j \partial x_j}, \tag{1.3}$$

where $i, j = 1, 2, 3$.

In accordance with suggestion of Reynolds, we represent the actual values of the parameters of turbulent flow as the sum of two components, namely,

$$\theta_i(\tau) = \Theta_i + \theta'_i(\tau), \tag{1.4}$$

where Θ_i is the time-averaged local value of this quantity, and θ'_i is its fluc-tuation value (the deviation of the instantaneous value from the local one).

Time averaging is performed as

$$\Theta_i = \frac{1}{T} \int_0^T \theta_i(\tau) d\tau. \tag{1.5}$$

Note that the averaging period T, on one hand, must exceed significantly the characteristic timescale of turbulent fluctuations and, on the other hand, must be much less than the characteristic time of variation of the macroscopic parameters of turbulent flow.

We subject (1.1)–(1.3) to the procedure of time averaging suggested by Reynolds to derive averaged equations of continuity, motion, and energy in the following form:

$$\sum_j \frac{\partial U_j}{\partial x_j} = 0, \tag{1.6}$$

$$\frac{\partial U_i}{\partial \tau} + \sum_j U_j \frac{\partial U_i}{\partial x_j} = -\frac{1}{\rho}\frac{\partial P}{\partial x_i} + \nu \sum_j \frac{\partial^2 U_i}{\partial x_j \partial x_j} - \sum_j \frac{\partial(\overline{u'_i u'_j})}{\partial x_j}, \tag{1.7}$$

$$\frac{\partial T}{\partial \tau} + \sum_j U_j \frac{\partial T}{\partial x_j} = a \sum_j \frac{\partial^2 T}{\partial x_j \partial x_j} - \sum_j \frac{\partial(\overline{u'_j t'})}{\partial x_j}. \tag{1.8}$$

Equations (1.7) and (1.8) were derived in view of the fact that

$$\sum_j \overline{u'_j \frac{\partial u'_i}{\partial x_j}} = \sum_j \frac{\partial(\overline{u'_i u'_j})}{\partial x_j}, \quad \sum_j \overline{u'_j \frac{\partial t'}{\partial x_j}} = \sum_j \frac{\partial(\overline{u'_j t'})}{\partial x_j}.$$

The validity of these relations may be demonstrated as

$$\sum_j \frac{\partial(\overline{u'_i u'_j})}{\partial x_j} = \sum_j \overline{u'_i \frac{\partial u'_j}{\partial x_j}} + \sum_j \overline{u'_j \frac{\partial u'_i}{\partial x_j}} = \overline{u'_i \sum_j \frac{\partial u'_j}{\partial x_j}} + \sum_j \overline{u'_j \frac{\partial u'_i}{\partial x_j}} = \sum_j \overline{u'_j \frac{\partial u'_i}{\partial x_j}}$$

$$\sum_j \frac{\partial(\overline{u'_j t'})}{\partial x_j} = \sum_j \overline{u'_j \frac{\partial t'}{\partial x_j}} + \sum_j \overline{t' \frac{\partial u'_j}{\partial x_j}} = \sum_j \overline{u'_j \frac{\partial t'}{\partial x_j}} + \overline{t' \sum_j \frac{\partial u'_j}{\partial x_j}} = \sum_j \overline{u'_j \frac{\partial t'}{\partial x_j}},$$

because

$$\overline{u'_i \sum_j \frac{\partial u'_j}{\partial x_j}} = 0, \quad \overline{t' \sum_j \frac{\partial u'_j}{\partial x_j}} = 0$$

by virtue of the fluctuation equation of continuity which will be derived below.

One can readily derive fluctuation equations of continuity, motion, and energy by subtracting (1.6)–(1.8) from (1.1)–(1.3), respectively,

$$\sum_j \frac{\partial u'_j}{\partial x_j} = 0, \tag{1.9}$$

$$\frac{\partial u'_i}{\partial \tau} + \sum_j \left[u'_j \frac{\partial U_i}{\partial x_j} + U_j \frac{\partial u'_i}{\partial x_j} + \frac{\partial (u'_i u'_j)}{\partial x_j} \right]$$

$$= -\frac{1}{\rho} \frac{\partial p'}{\partial x_i} + \nu \sum_j \frac{\partial^2 u'_i}{\partial x_j \partial x_j} + \sum_j \frac{\partial (\overline{u'_i u'_j})}{\partial x_j}, \tag{1.10}$$

$$\frac{\partial t'}{\partial \tau} + \sum_j \left[u'_j \frac{\partial T}{\partial x_j} + U_j \frac{\partial t'}{\partial x_j} + \frac{\partial (u'_j t')}{\partial x_j} \right]$$

$$= a \sum_j \frac{\partial^2 t'}{\partial x_j \partial x_j} + \sum_j \frac{\partial (\overline{u'_j t'})}{\partial x_j}. \tag{1.11}$$

For the majority of engineering calculations, it is sufficient to know the averaged parameters of gas, which may be determined by solving the averaged Navier–Stokes equations. However, unlike the case of laminar flow, the system of equations describing the averaged characteristics of turbulent flow (1.6)–(1.8) turns out to be open, because it contains unknown binary correlations in addition to the values of averaged velocity, temperature, and other thermodynamic parameters.

The transfer equation for Reynolds stresses $\overline{u'_i u'_j}$ is most often constructed as follows. First, we replace the subscripts j by k in (1.10) for u'_i and multiply both parts of the resultant equation by u'_j:

$$u'_j \frac{\partial u'_i}{\partial \tau} + \sum_k \left[u'_j u'_k \frac{\partial U_i}{\partial x_k} + u'_j U_k \frac{\partial u'_i}{\partial x_k} + u'_j \frac{\partial (u'_i u'_k)}{\partial x_k} \right]$$

$$= -u'_j \frac{1}{\rho} \frac{\partial p'}{\partial x_i} + u'_j \nu \sum_k \frac{\partial^2 u'_i}{\partial x_k \partial x_k} + u'_j \sum_k \frac{\partial (\overline{u'_i u'_k})}{\partial x_k}. \tag{1.12}$$

We write a similar equation for u'_j and multiply both its parts by u'_i:

$$u'_i \frac{\partial u'_j}{\partial \tau} + \sum_k \left[u'_i u'_k \frac{\partial U_j}{\partial x_k} + u'_i U_k \frac{\partial u'_j}{\partial x_k} + u'_i \frac{\partial (u'_j u'_k)}{\partial x_k} \right]$$

$$= -u'_i \frac{1}{\rho} \frac{\partial p'}{\partial x_j} + u'_i \nu \sum_k \frac{\partial^2 u'_j}{\partial x_k \partial x_k} + u'_i \sum_k \frac{\partial (\overline{u'_j u'_k})}{\partial x_k}. \tag{1.13}$$

We add (1.12) and (1.13) term by term and perform averaging. As a result, the transfer equation for Reynolds stresses assumes the form

$$\frac{\partial(\overline{u_i' u_j'})}{\partial \tau} + \sum_k U_k \frac{\partial(\overline{u_i' u_j'})}{\partial x_k} = \sum_k \frac{\partial}{\partial x_k}\left[\nu \frac{\partial(\overline{u_i' u_j'})}{\partial x_k} - \overline{u_i' u_j' u_k'}\right]$$

$$- \sum_k \left[(\overline{u_j' u_k'})\frac{\partial U_i}{\partial x_k} + (\overline{u_i' u_k'})\frac{\partial U_j}{\partial x_k}\right]$$

$$- \frac{1}{\rho}\left(\overline{u_i' \frac{\partial p'}{\partial x_j}} + \overline{u_j' \frac{\partial p'}{\partial x_i}}\right) - 2\nu \sum_k \overline{\frac{\partial u_i'}{\partial x_k}\frac{\partial u_j'}{\partial x_k}}. \quad (1.14)$$

In deriving (1.14), the following simple transformations were used:

$$\sum_k \frac{\partial(\overline{u_i' u_j' u_k'})}{\partial x_k} = \sum_k \overline{u_j' u_k' \frac{\partial u_i'}{\partial x_k}} + \sum_k \overline{u_i' \frac{\partial(u_j' u_k')}{\partial x_k}}$$

$$= \sum_k \overline{u_j' \frac{\partial(u_i' u_k')}{\partial x_k}} + \sum_k \overline{u_i' \frac{\partial(u_j' u_k')}{\partial x_k}},$$

$$\nu \sum_k \frac{\partial^2(\overline{u_i' u_j'})}{\partial x_k \partial x_k} - 2\nu \sum_k \overline{\frac{\partial u_i'}{\partial x_k}\frac{\partial u_j'}{\partial x_k}} = \nu \sum_k \overline{u_j' \frac{\partial^2 u_i'}{\partial x_k \partial x_k}} + \nu \sum_k \overline{u_i' \frac{\partial^2 u_j'}{\partial x_k \partial x_k}}.$$

The terms on the left-hand side of (1.14) describe the time variation and convection transfer of turbulent stresses, respectively. The terms on the right-hand side are responsible for the diffusion molecular and turbulent transfer, the generation of turbulent stresses from averaged motion, the fluctuation energy exchange between different components as a result of correlations of pressure fluctuations, and the viscous dissipation of turbulent energy, respectively. Equation (1.14) for the second moments contains unknown triple correlations; for the latter correlations, equations may be constructed which, in turn, contain the fourth moments. In order to derive a closed system of equations, the process of construction of equations must be interrupted at some stage. This is usually done by introducing additional hypotheses (models) for the correlation between the "higher" and "lower" moments (hypothesis of Millionshchikov and other hypotheses). Therefore, different models of turbulence are employed for closing the system of averaged Reynolds equations. Models of turbulent flows, which have gained the widest recognition in the literature, are briefly reviewed below.

1.2.1 Algebraic Models of Turbulence

When such models are employed, the double correlations appearing in equations are expressed in terms of averaged parameters. Relations between correlations and averaged characteristics of flow are often referred to as turbulence hypotheses. For example, the Boussinesq gradient approach has found extensive application. In accordance with this approach, the basic component of tensor $\overline{u_i' u_j'}$, i.e., the correlation $\overline{u_x' u_y'}$ responsible for the turbulent transfer of momentum, is represented as

$$\overline{u_x' u_y'} = -\nu_t \frac{\partial U_x}{\partial y}, \tag{1.15}$$

where ν_t is the turbulent analog of the coefficient of kinematic viscosity.

Note that the application of the Boussinesq hypothesis does not eliminate all difficulties, because it consists in fact in replacing one unknown $(\overline{u_x' u_y'})$ by another (ν_t). In order to determine the unknown ν_t, the semiempirical theory of turbulence (mixing length theory) of Prandtl is extensively used, according to which the fluctuations of longitudinal velocity are proportional to the gradient of averaged velocity. The values of longitudinal and transverse fluctuations of velocity must be close to each other in accordance with the continuity equation. Therefore, we have

$$u_y' \approx u_x' = l \frac{\partial U_x}{\partial y}, \tag{1.16}$$

where l is the mixing length characterizing the distance over which the turbulent moles (eddies) retain their individuality. It follows further from the continuity equation that the product $\overline{u_x' u_y'}$ must always be negative, because the fluctuations of velocities in different directions are opposite in sign. In view of the high correlation between the longitudinal and transverse fluctuations of velocity, we have

$$\overline{u_x' u_y'} = -l^2 \left(\frac{\partial U_x}{\partial y} \right)^2. \tag{1.17}$$

The coefficient ν_t may be determined from (1.15) and (1.17) as

$$\nu_t = l^2 \left| \frac{\partial U_x}{\partial y} \right|. \tag{1.18}$$

Note that the mixing length is not a universal quantity and assumes different values at different points of flow.

The disadvantage of the Prandtl theory, as well as of the existing varieties of algebraic models of turbulence (Karman, Van Driest, and others), is that it is based on the hypothesis of locality of the mechanism of turbulent transfer. According to this hypothesis, turbulent stresses depend only on the local structure of averaged flow. As a result, algebraic models of turbulence are capable of adequately describing only close-to-equilibrium flows. Additional "relaxation" terms including convection and diffusion terms must be introduced in order to calculate substantially nonequilibrium flows. Neither may the algebraic (local) models of turbulence be used to calculate turbulent flows beyond the boundary layer, which are characterized by highly uniform averaged velocities (nonshear turbulent flows).

In recent decades, more sophisticated (as regards their physical content) differential models of turbulence find extensive application for the calculation of turbulent single-phase flows. In addition to equations for averaged quantities, such models include additional differential equations of transfer for the most important characteristics of the structure of turbulence. The differential models are divided into one parameter, two parameter, and so on by the number of additional (to the averaged ones) equations.

1.2.2 One-Parameter Models of Turbulence

Models of turbulence of this class are based on the use of the transfer equation for determining one of the characteristics of turbulence. Such a characteristic is most frequently provided by the turbulent energy k, turbulent viscosity ν_t, or by the basic component of stress tensor $\overline{u'_x u'_y}$.

Model based on the equation for turbulent energy

In order to derive the transfer equation for turbulent energy, (1.10) for fluctuation motion must be multiplied by u'_i, summed with respect to i, and then averaged. The resultant equation will have the form

$$\frac{\partial k}{\partial \tau} + \sum_j U_j \frac{\partial k}{\partial x_j} = \sum_j \frac{\partial}{\partial x_j} \left[\nu \frac{\partial k}{\partial x_j} - \overline{u'_j \left(\frac{1}{2} \sum_i u_i'^2 + \frac{p'}{\rho} \right)} \right]$$

$$- \sum_j \sum_i \overline{u'_i u'_j} \frac{\partial U_i}{\partial x_j} - \nu \sum_j \sum_i \overline{\frac{\partial u'_i}{\partial x_j} \frac{\partial u'_i}{\partial x_j}}, \qquad (1.19)$$

where $k = \frac{1}{2} \sum_i \overline{u_i'^2}$.

Equation (1.19) was first derived by Kolmogorov [13]. The terms on the left-hand side of (1.19) describe the time variation and convection transfer of turbulent energy, respectively. The first term on the right-hand side describes the diffusion of turbulent energy, the second term, its generation owing to the energy of averaged motion, and the third term, the dissipation due to viscosity.

The equation for k is not closed by virtue of indeterminacy of all terms on the right-hand side, namely, the diffusion, generation, and dissipation terms. The following hypotheses of Kolmogorov [13] are usually employed to find the diffusion and generation terms:

$$-\overline{u'_j\left(\frac{1}{2}\sum_i u'^2_i + \frac{p'}{\rho}\right)} = \frac{\nu_t}{\sigma_k}\frac{\partial k}{\partial x_j}, \tag{1.20}$$

$$-\overline{u'_i u'_j} = \nu_t\left(\frac{\partial U_i}{\partial x_j} + \frac{\partial U_j}{\partial x_i}\right) - \frac{2}{3}k\delta_{ij}, \tag{1.21}$$

where σ_k is an empirical constant (it is most frequently assumed that $\sigma_k = 1$), and δ_{ij} is the Kronecker symbol.

Within one-parameter models, the unknown term responsible for the dissipation of energy due to viscosity ε is usually represented in terms of the turbulent energy k and its scale l in the form

$$\varepsilon = \nu\sum_j\sum_i \overline{\frac{\partial u'_i}{\partial x_j}\frac{\partial u'_i}{\partial x_j}} = C_\mu^{3/4}\frac{k^{3/2}}{l}, \tag{1.22}$$

with the most frequently employed value of the coefficient $C_\mu = 0.09$.

Therefore, (1.19) for turbulent energy may be rewritten as

$$\frac{\partial k}{\partial \tau} + \sum_j U_j\frac{\partial k}{\partial x_j} = \sum_j\frac{\partial}{\partial x_j}\left[\left(\nu + \frac{\nu_t}{\sigma_k}\right)\frac{\partial k}{\partial x_j}\right]$$
$$+ \sum_j\sum_i\left[\nu_t\left(\frac{\partial U_i}{\partial x_j} + \frac{\partial U_j}{\partial x_i}\right) - \frac{2}{3}k\delta_{ij}\right]\frac{\partial U_i}{\partial x_j} - \varepsilon. \tag{1.23}$$

We denote the diffusion and generation terms by D and P, respectively; then, (1.23) may be written in a condensed form

$$\frac{Dk}{D\tau} = D + P - \varepsilon \tag{1.24}$$

The turbulent viscosity is expressed in terms of turbulent energy as follows [13]

$$\nu_t = C_\nu k^{1/2}l. \tag{1.25}$$

Therefore, the set of equations (1.6), (1.7), and (1.23) in view of relations (1.22) and (1.25) gives a closed description of momentum transfer in a single-phase turbulent flow.

1.2.3 Two-Parameter Models of Turbulence

Models of turbulence of this class contain as many as two differential equations for the characteristics of turbulence. As a rule, the first equation in two-parameter models is the equation for turbulent energy derived and analyzed earlier. The most universally employed model is the two-parameter $k-\varepsilon$ model of turbulence, in which the equation for the rate of dissipation of turbulent fluctuations is used as the second equation.

The transfer equation for turbulent dissipation, which was first derived by Davydov [6], has the form

$$
\frac{\partial \varepsilon}{\partial \tau} + \sum_j U_j \frac{\partial \varepsilon}{\partial x_j} = -\sum_j \frac{\partial}{\partial x_j} \left(\overline{u'_j \sum_{i,k} \nu \frac{\partial u'_i}{\partial x_k} \frac{\partial u'_i}{\partial x_k}} + \frac{2\nu}{\rho} \sum_k \overline{\frac{\partial p'}{\partial x_k} \frac{\partial u'_j}{\partial x_k}} \right)
$$

$$
+ \nu \sum_j \frac{\partial^2 \varepsilon}{\partial x_j^2} - 2\nu \sum_{j,i,k} \frac{\partial U_i}{\partial x_j} \left(\overline{\frac{\partial u'_k}{\partial x_i} \frac{\partial u'_k}{\partial x_j}} + \overline{\frac{\partial u'_i}{\partial x_k} \frac{\partial u'_j}{\partial x_k}} \right)
$$

$$
- 2\nu \sum_{j,i,k} \overline{\frac{\partial u'_i}{\partial x_k} \frac{\partial u'_i}{\partial x_j} \frac{\partial u'_j}{\partial x_k}} - 2\nu^2 \sum_{j,i,k} \overline{\left(\frac{\partial^2 u'_i}{\partial x_j \partial x_k} \right)^2}
$$

$$
- 2\nu \sum_{j,i,k} \frac{\partial^2 U_i}{\partial x_j \partial x_k} \overline{u'_j \frac{\partial u'_i}{\partial x_k}}, \tag{1.26}
$$

where

$$
\varepsilon = \nu \sum_j \sum_i \overline{\frac{\partial u'_i}{\partial x_j} \frac{\partial u'_i}{\partial x_j}}.
$$

We now describe the physical meaning of the terms appearing in (1.26). The first and second terms on the left-hand side characterize the time variation and convection transfer of dissipation of turbulent energy, respectively. The first and second terms on the right-hand side characterize the turbulent and molecular diffusion of dissipation of turbulent energy. The third term represents the generation of dissipation. The fourth and fifth terms characterize the decrease in dissipation under the effect of viscosity and due to turbulent deformation. The sixth term is usually ignored (as well as the fourth one).

Within the gradient approximations for the diffusion and generation terms, (1.26) is usually represented in the following form [6, 14]

$$
\frac{\partial \varepsilon}{\partial \tau} + \sum_j U_j \frac{\partial \varepsilon}{\partial x_j} = \sum_j \frac{\partial}{\partial x_j} \left[\left(\nu + \frac{\nu_t}{\sigma_\varepsilon} \right) \frac{\partial \varepsilon}{\partial x_j} \right]
$$

$$
- C_{\varepsilon 1} \frac{\varepsilon}{k} \sum_j \sum_i \overline{u'_i u'_j} \frac{\partial U_i}{\partial x_j} - C_{\varepsilon 2} \frac{\varepsilon^2}{k}, \tag{1.27}
$$

where $\sigma_\varepsilon, C_{\varepsilon 1}$, and $C_{\varepsilon 2}$ are empirical constants. The universally accepted values of these constants are $\sigma_\varepsilon = 1.3$, $C_{\varepsilon 1} = 1.44$, and $C_{\varepsilon 2} = 1.92$.

Note that, in calculating wall-bounded flows, some "wall" functions are introduced for determining the terms on the right-hand side of (1.27). These correction functions are intended to ensure that the behavior of the quantities determined as a result of calculations should fit the available experimental data.

One can readily see that the form of the equation for dissipation agrees with that of the equation for turbulent energy and may also be represented in a condensed form

$$\frac{D\varepsilon}{D\tau} = D_\varepsilon + P_\varepsilon - \varepsilon_\varepsilon, \tag{1.28}$$

where D_ε, P_ε, and ε_ε denote the diffusion, generation, and decrease of the dissipation of turbulent energy, respectively.

The correlation between the turbulent energy and the rate of its dissipation is determined using the Kolmogorov–Prandtl relation

$$\nu_t = C_\mu \frac{k^2}{\varepsilon}, \quad C_\mu = 0.09. \tag{1.29}$$

Therefore, the set of equations (1.6), (1.7), (1.23), and (1.27) in view of relation (1.29) gives a closed description of momentum transfer in a single-phase turbulent flow.

A number of other models of single-phase turbulent flows exist in addition to those described earlier. In my opinion, quite a successful comparison of different differential models for the calculation of wall flows was made by Lushchik and Yakubenko [16].

1.3 Main Characteristics of Single-Phase Flows

In this section, a brief look is taken at the main characteristics of turbulent flows of gas and their distribution over the channel (pipe) cross-section. Such characteristics include the averaged and fluctuation velocities, the turbulent energy and its spectrum, and the space and timescales of turbulence.

1.3.1 Distributions of Averaged Velocity

The turbulent flow is the most complex and most commonly occurring form of flow of continuous medium. The transition from eddy-free laminar to turbulent flow occurs as a result of loss of hydrodynamic stability which occurs when some critical value of dimensionless parameter (Reynolds number) is attained. The Reynolds number for a round pipe flow has the form

$$Re_D = \frac{\langle U_x \rangle 2R}{\nu}, \tag{1.30}$$

where $\langle U_x \rangle = \frac{2}{R^2} \int_0^R U_x(r) r \, dr$ is the cross-section average velocity of continuous medium, R is the pipe radius, and ν is the coefficient of kinematic viscosity of continuous medium.

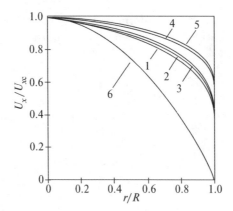

Fig. 1.1. The distribution of averaged velocities under conditions of turbulent pipe flow of air: (1) n = 6, (2) n = 7, (3) n = 8.8, (4) n = 9.8, (5) n = 10, (6) parabolic laminar profile

The stabilized profile of averaged velocity in channels (pipes) in the flow core is often described by the power law

$$\frac{U_x}{U_{xc}} = \left(\frac{R - r}{R}\right)^{1/n},\tag{1.31}$$

where r is the distance from the pipe axis, and n is a function of the Reynolds number.

For a developed turbulent pipe flow, the exponent in power law (1.31) assumes the following values (Fig. 1.1): $n = 6$ for $Re_D = 4\times10^3$, $n = 7$ for $Re_D = 1.1\times10^5$, $n = 8.8$ for $Re_D = 1.1\times10^6$, $n = 9.8$ for $Re_D = 2\times10^6$, and $n = 10$ for $Re_D = (2-3.2)\times10^6$. The shape of averaged velocity profile is further affected by the degree of turbulence. An increase in the degree of turbulence causes the velocity profile to become flatter owing to the velocity increase in the vicinity of the wall, which leads to an increase in n.

The description of averaged velocity distribution may further involve the use of universal coordinates, i.e., the form $U^+ = U^+(y^+)$, where $U^+ = U_x/u_*$ and $y^+ = yu_*/\nu$. The profile of averaged velocity of stabilized flow is described by the relation

$$U^+ = \frac{1}{\gamma}\ln y^+ + C + \frac{1}{\gamma}\Pi\omega,\tag{1.32}$$

where γ and C are constants, Π is the wake parameter, and ω is the wake function.

In the wall region ($y^+ > 30$) of developed turbulent flow ($Re_D > 10^4$), the last term of (1.32) may be ignored, and this equation transforms to the so-called "logarithmic law of the wall,"

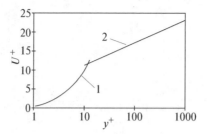

Fig. 1.2. The distribution of averaged velocity in universal coordinates: (1) viscous sublayer and (2) turbulent core

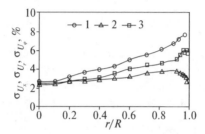

Fig. 1.3. The distribution of three components of the intensity of fluctuations of air velocity in the pipe cross section: (1) σ_{U_x}, (2) σ_{U_r}, (3) σ_{U_φ}

$$U^+ = \frac{1}{\gamma}\ln y^+ + C, \tag{1.33}$$

in which the constants are most frequently taken to be $\gamma = 0.4$ and $C = 5.5$.

In the viscous sublayer region ($y^+ < 10$), (1.33) transforms to the linear dependence

$$U^+ = y^+ \tag{1.34}$$

The distribution of the velocity of stabilized pipe flow in universal coordinates is given in Fig. 1.2.

1.3.2 Distributions of Averaged Fluctuation Velocities

Figure 1.3 gives the classical data of Laufer on the distribution of three components of relative intensity of turbulence $\sigma_{U_x} = (\overline{u_x'^2})^{1/2}/U_{xc}, \sigma_{U_r} = (\overline{u_r'^2})^{1/2}/U_{xc}$, and $\sigma_{U_\varphi} = (\overline{u_\varphi'^2})^{1/2}/U_{xc}$ under conditions of pipe flow. The experiments were performed with a pipe of inside diameter $D = 250\,\text{mm}$. The cited results relate to $U_{xc} \approx 3\,\text{ms}^{-1}$, which corresponds to the value of Reynolds number $Re_D = 50{,}000$. One can infer from Fig. 1.3 that the distributions of all components of fluctuation velocity are substantially nonuniform. The isotropic conditions are attained in a region which is far removed from the pipe wall. The intensity of turbulence in the axial direction exceeds the respective characteristics in the normal and azimuthal directions.

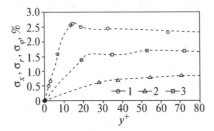

Fig. 1.4. The distribution of three components of the intensity of fluctuations, related to dynamic velocity, in the pipe wall region: (1) σ_x, (2) σ_r, (3) σ_φ

The axial component of fluctuations has a maximum which is reached in a region that is very close to the wall (see Fig. 1.4). The coordinate of this maximum is $y^+ \approx 15$. The distribution of normal fluctuation of velocity also exhibits a maximum located much farther away from the wall ($r/R \approx 0.9$). By its magnitude, the azimuthal component of velocity fluctuations takes the midposition between the axial and normal components over the entire pipe crosssection. The distribution of the components of intensity of fluctuations $\sigma_x = (\overline{u_x'^2})^{1/2}/u_*, \sigma_r = (\overline{u_r'^2})^{1/2}/u_*$, and $\sigma_\varphi = (\overline{u_\varphi'^2})^{1/2}/u_*$, related to dynamic velocity, is given in Fig. 1.4. These data indicate that, in the vicinity of the wall, the transverse component is close in magnitude to the dynamic velocity, and the axial component is more than twice the dynamic velocity.

1.3.3 Turbulent Energy

The fluctuation velocity profiles obtained by Laufer were used to find the turbulent energy of air. The calculated distribution of kinetic turbulent energy k/u_*^2 related to the square of dynamic velocity over the pipe cross-section is given in Fig. 1.5.

Following are some important remarks concerning the turbulent energy balance. In the region closest to the wall ($y^+ \leq 20$; $r/R \geq 0.998$), the terms which characterize the generation and dissipation of turbulent energy are in fact equal to each other but opposite in sign. The same is true of the terms responsible for the turbulent diffusion of kinetic energy and diffusion of energy of pressure fluctuations. In the region farthest from the wall ($y^+ > 20$; $0.7 < r/R < 0.998$), the main contribution to the turbulent energy balance is made by the generation and dissipation, with preservation of insignificant energy transfer due to turbulent diffusion (diffusion of kinetic energy and energy of pressure fluctuations). Away from the pipe wall, the transfer of energy of pressure fluctuations decreases to become negligible in the central region ($r/R < 0.4$). At the same time, the importance of turbulent diffusion of kinetic energy in the overall balance continues to increase, because the generation and dissipation of turbulence decrease. In the region close to the pipe axis ($r/R < 0.2$), the generation of turbulence is close to zero,

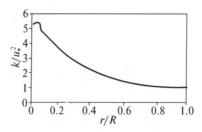

Fig. 1.5. The distribution of turbulent energy under conditions of pipe flow

and the dissipation of energy is balanced only by turbulent diffusion of kinetic energy.

1.3.4 Energy Spectrum of Turbulence

Fluctuations of physical quantities in turbulent flows occur at most diverse frequencies; therefore, the dispersion (or mean-square deviation) of fluctuating quantity may be represented as the sum of respective quantities relating to different frequencies. Such distribution of a fluctuating quantity is an energy spectrum

$$E_i(f) = \frac{d\overline{\theta_i'^2}}{df},\tag{1.35}$$

where $E_i(f)$ defines the fraction of dispersion $d\overline{\theta_i'^2}$ of some fluctuating quantity θ_i, which corresponds to the frequency band df. In doing so, one must bear in mind that

$$\overline{\theta_i'^2} = \int_0^\infty E_i(f)df.\tag{1.36}$$

For convenience in representing energy spectra, they are usually normalized to the value of dispersion of the quantity being treated, namely,

$$\overline{E}_i(f) = \frac{E_i(f)}{\overline{\theta_i'^2}},\tag{1.37}$$

where $\int_0^\infty \overline{E}_i(f)df$ is the normalization condition.

The energy spectra of the components of velocity fluctuations $\overline{E}_x(f), \overline{E}_r(f)$ and $\overline{E}_\varphi(f)$ under conditions of pipe flow, obtained experimentally by Laufer, are given in Fig. 1.6.

1.3.5 Correlations in Turbulent Flows

The need often arises for finding the statistical correlation between fluctuating quantities (characteristics) of turbulent flow. The measure of such correlation between quantities is provided by correlation coefficients which represent the

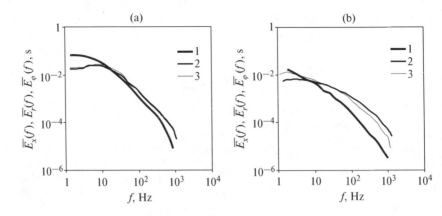

Fig. 1.6. Energy spectrum of velocity fluctuations (**a**) on the axis (r/R $= 0$) and (**b**) in the wall region (r/R $= 0.926$) of the pipe: (1) $\overline{E}_x(f)$, (2) $\overline{E}_r(f)$, (3) $\overline{E}_\varphi(f)$

ratio of the averaged product of two instantaneous values of given fluctuating parameters to the product of their averaged values. A large number of correlation coefficients of diverse forms exist, of which the most extensively employed ones are one-point, two-point, and Eulerian time coefficients.

One-Point Correlations

The one-point coefficient of correlation of two different fluctuating quantities defines their correlation at a given point of space at one and the same instant of time. This correlation coefficient is expressed as

$$R_{i,j} = \frac{\overline{\theta_i' \theta_j'}}{(\overline{\theta_i'^2})^{1/2}(\overline{\theta_j'^2})^{1/2}}, \tag{1.38}$$

where θ_i' and θ_j' and denote the fluctuation values of θ_i and θ_j.

Two-Point Correlations

The two-point coefficient of correlation of a fluctuating quantity defines the correlation of its values at two different points of space at one and the same instant of time and is written as

$$R_{i,x} = \frac{\overline{\theta_{i,x_1}' \theta_{i,x_2}'}}{(\overline{\theta_{i,x_1}'^2})^{1/2}(\overline{\theta_{i,x_2}'^2})^{1/2}}, \tag{1.39}$$

where θ_{i,x_1}' and θ_{i,x_2}' denote the fluctuation values of θ_i at points of space with coordinates $x = x_1$ and $x = x_2$.

Eulerian Time Correlation

The coefficient of Eulerian time autocorrelation defines the internal correlation of a fluctuating quantity at different instants of time at a given point of space and has the form

$$R_{i,\tau} = \frac{\overline{\theta'_{i,\tau_1}\,\theta'_{i,\tau_2}}}{\overline{\theta'^2_i}}, \tag{1.40}$$

where θ'_{i,τ_1} and θ'_{i,τ_2} denote the fluctuation values of θ_i at instants of time $\tau = \tau_1$ and $\tau = \tau_2$.

Correlations are present in equations of turbulent flows and are used extensively in theoretical analysis.

1.3.6 Scales of Turbulent Flows

Time and space scales of flows are recognized. The most frequently employed scales of turbulent flows are considered below.

Characteristic Scales of Gas in Averaged Motion

The turbulent flows treated in this monograph are steady state on the average. Nevertheless, it would be useful to introduce some characteristic timescale of the carrier phase, which is necessary for analysis of the process of relaxation of averaged velocities of gas and particles. We define this scale as

$$T_f = \frac{L}{U_x}, \tag{1.41}$$

where L is some characteristic length over which the process of relaxation of phase velocities occurs, and U_x is the averaged velocity of gas.

The characteristic geometric dimension may be provided by the length over which the particles are accelerated from the point of their injection into the flow to the pipe cross-section of interest to us, by the distance from the critical point upstream of the body subjected to flow to the point at which the deceleration of gas begins, by the distance from the critical point downstream of the body along its surface to the cross-section in the boundary layer of interest to us, and so on.

The characteristic averaged velocity of gas may be selected from the velocity of flow on the pipe axis, the velocity of flow undisturbed by the presence of a body, and the velocity on the external bound of the boundary layer in analyzing heterogeneous flows in pipes, under conditions of flow past bodies, and in the boundary layer, respectively.

Characteristic Scales of Gas in Large-Scale Fluctuation Motion

Eulerian and Lagrangian scales of flows are recognized. The integral Eulerian time and space scales of turbulence are most commonly determined in terms of the coefficient of Eulerian time autocorrelation as follows:

$$T_{\rm E} = \int_0^\infty R_{x,\tau}(\tau)d\tau, \tag{1.42}$$

$$L_{\rm E} = U_x \int_0^\infty R_{x,\tau}(\tau)d\tau = U_x T_{\rm E}. \tag{1.43}$$

These scales characterize the time, over which one still observes a marked correlation between fluctuations of gas velocity (the lifetime of large turbulent eddies), and the size of energy-carrying eddies, respectively.

Note that the correlation $R_{x,\tau}(\tau)$ may also be measured in coordinates moving at the average velocity of flow. We will designate this correlation as $R_{x,\tau}^0(\tau)$. In this case, we have for the integral timescale in moving coordinates $T_{\rm E}^0$

$$T_{\rm E}^0 = \int_0^\infty R_{x,\tau}^0(\tau)d\tau. \tag{1.44}$$

The integral Eulerian space scale of turbulence will be represented as

$$L_{\rm E} \approx (\overline{u_x'^2})^{1/2} T_{\rm E}^0. \tag{1.45}$$

We use (1.44) and (1.45) to derive the following correlation between Eulerian timescales of turbulence measured in different coordinates:

$$\frac{T_{\rm E}^0}{T_{\rm E}} \approx \frac{U_x}{(\overline{u_x'^2})^{1/2}}. \tag{1.46}$$

In analyzing flows of gas with particles, the Lagrangian time $T_{\rm L}$ and space $L_{\rm L}$ scales of turbulence are used most frequently. The available literature data on the ratio between the Eulerian and Lagrangian scales of turbulence are contradictory. As a first approximation, we can assume that

$$L_{\rm L} \approx L_{\rm E}, \tag{1.47}$$

$$T_{\rm L} \approx T_{\rm E}^0. \tag{1.48}$$

The following relation is frequently used to determine the integral timescale of turbulence:

$$T_{\rm L} = C_\mu^{1/2} \frac{k}{\varepsilon}. \tag{1.49}$$

As applied to pipe flow, the rate of dissipation of turbulent energy ε is determined as

$$\varepsilon = C_\mu^{3/4} \frac{k^{3/2}}{l}, \tag{1.50}$$

where the Prandtl–Nikuradse mixing length l has the form

$$l = 0.4y(1 - 1.1\bar{y} + 0.6\bar{y}^2 - 0.15\bar{y}^3), \quad \bar{y} = y/R. \tag{1.51}$$

Characteristics Scales of Gas in Small-Scale Fluctuation Motion

The Eulerian time microscales of turbulence in fixed and moving coordinates are determined as follows:

$$\tau_E = 2\left[-\left(\frac{\mathrm{d}^2 R_{x,\tau}(\tau)}{\mathrm{d}\tau^2}\right)_{\tau=0}\right]^{-1/2}, \tag{1.52}$$

$$\tau_E^0 = 2\left[-\left(\frac{\mathrm{d}^2 R_{x,\tau}^0(\tau)}{\mathrm{d}\tau^2}\right)_{\tau=0}\right]^{-1/2}. \tag{1.53}$$

These scales define the peak width of the $R_{x,\tau}(\tau)$ and $R_{x,\tau}^0(\tau)$ curves in the vicinity of point $\tau = 0$ and the lifetime of the smallest eddies causing the dissipation of kinetic energy to heat.

The time and space scales of the smallest dissipative eddies (microscales of turbulence) were introduced by Kolmogorov and are determined as

$$\tau_K = \left(\frac{\nu}{\varepsilon}\right)^{1/2}. \tag{1.54}$$

$$\eta = \left(\frac{\nu^3}{\varepsilon}\right)^{1/4}. \tag{1.55}$$

The space and timescales of flows, described above, are universally used in analyzing turbulent flows; they form the basis of the classification of heterogeneous flows developed in Sect. 1.5

1.4 Main Characteristics of Heterogeneous Flows

In addition to the characteristics of single-phase flows, heterogeneous flows exhibit a number of peculiar characteristics of their own. These characteristics of heterogeneous flows may be conventionally divided into intensive and extensive physical quantities. The intensive quantities include physical properties of particles such as their size (diameter) d_p and physical density ρ_p. In the case of nonisothermal flow, serious importance is assumed also by the heat capacity of the particle material C_{p_p}. The above-mentioned properties of particles characterize the dynamic and thermal inertia of the dispersed phase.

1.4.1 Time of Dynamic Relaxation of Particles

The dynamic inertia of particles is defined by the time of their relaxation τ_p which has the form

$$\tau_p = \frac{\tau_{p0}}{C} = \frac{\rho_p d_p^2}{18\mu C}, \tag{1.56}$$

where

$$C = \begin{cases} 1 + Re_p^{2/3}/6 & \text{at} \quad Re_p \leq 10^3, \\ 0.11 Re_p/6 & \text{at} \quad Re_p > 10^3. \end{cases}$$

In expression (1.56), τ_{p0} characterizes the time of dynamic relaxation of a Stokesian particle ($Re_p < 1$). Note that the inertia of the Stokesian particle further depends on the characteristics of the medium in which this particle moves. For example, the expression for τ_{p0} includes the coefficient of dynamic viscosity of the medium. The correction function C takes into account the effect of inertial forces on the time of relaxation of a non- Stokesian particle. Therefore, in the case of motion of the non-Stokesian particle, its inertia depends also on the dimensionless complex such as the Reynolds number of the particle calculated by the relative velocity between phases and the diameter of disperse impurity, i.e.,

$$Re_p = \frac{|\vec{W}| d_p}{\nu} = \frac{|\vec{U} - \vec{V}| d_p}{\nu}. \tag{1.57}$$

1.4.2 Time of Thermal Relaxation of Particles

The thermal inertia of particles is characterized by the time of their thermal relaxation τ_t determined by the relation

$$\tau_t = \frac{\tau_{t0}}{C_1} = \frac{C_{p_p} \rho_p d_p^2}{12 \lambda C_1} \tag{1.58}$$

where $C_1 = 1 + 0.3 Re_p^{1/2} Pr^{1/3}$.

In expression (1.58), τ_{t0} characterizes the time of thermal relaxation of a Stokesian particle ($Re_p < 1$). Note that the thermal inertia of the Stokesian particle (as does the dynamic inertia) depends both on its physical properties and heat capacity and on the thermal conductivity λ of the surrounding gas. The correction function C_1 takes into account the effect of convection on heat transfer between a non-Stokesian particle and surrounding gas. Therefore, the thermal inertia of the non-Stokesian particle depends also on the relative velocity between phases and on the viscosity and heat capacity of the carrier continuum.

1.4.3 Stokes Numbers

In this monograph, turbulent flows of gas with solid particles are treated. A turbulent flow is characterized by a number of space and corresponding timescales (see Sect. 1.3). As a result, it appears advisable to construct a number of dimensionless parameters, i.e., Stokes numbers which characterize the particle inertia with respect to some or other scales of flow.

In the case of particle motion in a flow of gas with a gradient of averaged velocity in the longitudinal direction (for example, under conditions of flow in nozzles and in the boundary layer or in the vicinity of bodies subjected to flow), as well as during acceleration of particles in a flow with a constant value of averaged velocity, one must take into account the inertia of particles when analyzing the process of relaxation of averaged velocities of phases. For this purpose, it is necessary to introduce the Stokes number in averaged motion, which will be written as

$$Stk_f = \frac{\tau_p}{T_f}, \tag{1.59}$$

where T_f is the characteristic time of the carrier phase in averaged motion.

The parameter of dynamic inertia of particles in large-scale fluctuation motion is represented by the Stokes number

$$Stk_L = \frac{\tau_p}{T_L}, \tag{1.60}$$

where T_L is the characteristic time of the carrier phase in large-scale fluctuation motion (integral Lagrangian timescale of turbulence).

The particle inertia in small-scale fluctuation motion will be likewise characterized by the Stokes number represented as

$$Stk_K = \frac{\tau_p}{\tau_K}, \tag{1.61}$$

where τ_K is the Kolmogorov timescale of turbulence.

Note that, in treating nonisothermal heterogeneous flows, it is necessary to introduce the appropriate dimensionless parameters which characterize the thermal inertia of particles relative to the respective characteristic timescales of variation of temperature of the carrier medium.

1.4.4 Particle Concentration

The particle concentration is an extensive physical characteristic of heterogeneous flows. Three types of concentration of the dispersed phase are recognized, namely, mass concentration, volume concentration, and number density.

We will consider two possible approaches to determining the particle concentration in the flow. According to the first approach, the local volume and mass concentration of particles is determined as follows (see Fig. 1.7):

$$\Phi = \frac{V_{p\Sigma}}{V_\Sigma} = \frac{V_{p\Sigma}}{V_g + V_{p\Sigma}}, \tag{1.62}$$

$$M = \frac{M_{p\Sigma}}{M_\Sigma} = \frac{M_{p\Sigma}}{M_g + M_{p\Sigma}}, \tag{1.63}$$

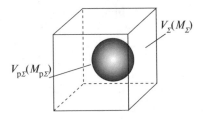

Fig. 1.7. Determination of the mass and volume concentrations of particles in the flow

where $V_{p\Sigma}$, V_g, and V_Σ denote the volumes taken up by particles and gas in the elementary volume of flow and the value of this volume, respectively; and $M_{p\Sigma}$, M_g, and M_Σ denote the mass of particles, the mass of gas in the elementary volume of flow, and the total mass of the elementary volume of heterogeneous flow, respectively. The values of volume and mass concentration of particles, determined by relations (1.62) and (1.63), are in the range from zero to unity.

In accordance with the second approach, the volume (mass) concentration of particles is determined as the ratio of the total volume (mass) of particles to the respective values for gas, i.e.,

$$\Phi = \frac{V_{p\Sigma}}{V_g}, \tag{1.64}$$

$$M = \frac{M_{p\Sigma}}{M_g}, \tag{1.65}$$

In contrast to relations (1.62) and (1.63), the volume and mass concentrations of particles determined by (1.64) and (1.65) may assume values from zero to infinity. In analyzing (1.62)–(1.65), it may be inferred that, for the case of heterogeneous flow with a low volume (mass) concentration of the dispersed phase, the values of the volume (mass) concentration of particles calculated by different relations will be close to one another.

The flows of gases with solid particles treated in this monograph are characterized by a relatively low volume concentration of particles. At the same time, the total mass of particles may be several times that of gas. Therefore, for the flows under study, the use of any of the foregoing relations for determining the volume concentration of particles will produce virtually the same results. As to the mass concentration of particles, its value according to (1.63) for flows in which the mass of particles is several times that of gas hardly varies with increasing content of the dispersed phase and tends asymptotically to unity, which is not quite convenient. Because of this, it is preferable to use expression (1.65) for determining the mass concentration of particles.

The use of relations (1.64) and (1.65) leads to obvious expressions relating the mass and volume concentrations and the number density of particles in the form

$$M = \frac{\Phi \rho_p}{\rho}, \tag{1.66}$$

$$\Phi = \frac{N \pi d_p^3}{6}. \tag{1.67}$$

The experimental investigations of heterogeneous flows often involve the use of the concepts of volume and mass flow rate concentrations, which we determine as follows for flows of the types being treated:

$$\Phi_G = \frac{G_{Vp}}{G_{Vg}}, \tag{1.68}$$

$$M_G = \frac{G_{Mp}}{G_{Mg}}, \tag{1.69}$$

where the volumetric and mass flow rates of particles (G_{Vp}, G_{Mp}) and gas (G_{Vg}, G_{Mg}) through some site S are given as

$$G_{Vp} = N \frac{\pi d_p^3}{6} V S, \tag{1.70}$$

$$G_{Vg} = U S, \tag{1.71}$$

$$G_{Mp} = N \rho_p \frac{\pi d_p^3}{6} V S, \tag{1.72}$$

$$G_{Mg} = \rho U S, \tag{1.73}$$

where N is the particle number density (the number of particles per unit volume), V is the normal (with respect to site S) velocity of particles, and U is the normal velocity of gas.

From (1.64)–(1.73), one can readily derive the expression for the correlation between the "true" values of volume (mass) concentration of the dispersed phase and the volume (mass) flow rate concentration,

$$\frac{\Phi}{\Phi_G} = \frac{M}{M_G} = \frac{U}{V}. \tag{1.74}$$

Therefore, for the cases of upward ($U > V$) and downward ($U < V$) flows which are not bounded by walls, we have $M > M_G$ and $M < M_G$, respectively.

In the foregoing, we talked about local concentrations of particles in the flow. In the case where the distribution of the dispersed phase i nonuniform over the channel (pipe) cross-section, i.e., $\Phi(r) \neq$ const. and $M(r) \neq$ const., it is often necessary to know the average (with respect to space) concentrations of particles, which are determined for pipe flow as

$$< \varPhi > = \frac{2}{R^2} \int_0^R \varPhi(r) r \mathrm{d}r,$$ (1.75)

$$< M > = \frac{2}{R^2} \int_0^R M(r) r \mathrm{d}r.$$ (1.76)

The foregoing main characteristics of heterogeneous flows identified above will be often employed in the presentation below and used in constructing the classification of turbulent flows of gas with solid particles.

1.5 Classification of Heterogeneous Turbulent Flows

In spite of numerous available monographs (see the foreword) dealing with the most diverse aspects of multiphase flows, no classification of turbulent hetero-geneous flows exists to date. The presence of numerous modes of flow of gas suspension, which are defined both by the parameters of carrier gas (physi-cal properties, Reynolds number, intensity of turbulent fluctuations, scales of turbulence, and so on) and by the parameters of particles proper (physical properties, Reynolds number of a particle, local concentration, polydisper-sion, and so on), complicates significantly the use of the classical modeling theory; this makes impossible the systematization and generalization of the investigation results. Attempts at systematizing heterogeneous flows by way of determining the ranges of validity of various numerical models [1, 5, 8, 22] compiling schemes of flow modes [18], and searching for a single universal parameter [10, 11, 21] defining the type of flow have failed, and the obtained classifications may hardly be regarded as comprehensive and complete. At the same time, the demand for the classification of such flows is extremely great.

I suggest determining the form (type) of heterogeneous flow by using the combination of classifications of two-phase flows with respect to the volume concentration and Stokes numbers (in averaged, large-scale, and small-scale fluctuation motion). It appears that it is only in this manner that one can estimate in advance the presence and intensity of basic interphase interactions and exchange processes.

Figure 1.8 gives possible varieties of particle-laden flows depending on their volume concentration [8]. If the volume concentration of disperse impurity is insignificant ($\varPhi \leq 10^{-6}$), its time average effect on the flow of the carrier medium is negligible. In heterogeneous flows of this type, the determining interaction is represented by the effect of the carrier phase on suspended par-ticles, which defines fully all of their characteristics (averaged and fluctuation velocities and temperatures, local concentration, etc.). When the volume con-centration increases ($10^{-6} < \varPhi \leq 10^{-3}$), the disperse impurity in turn begins to affect inversely the carrier medium. Heterogeneous flows of these two types are often referred to as dilute flows. In the case of a dense flow ($\varPhi > 10^{-3}$),

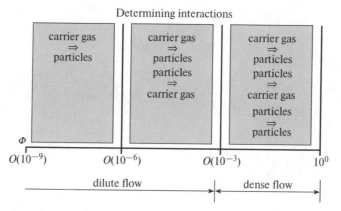

Fig. 1.8. Classification of heterogeneneous flows by the volume concentration of particles

the interaction of particles with one another is added to the already described interactions between suspended particles and the carrier phase.

Table 1.1 gives an indication of possible forms of turbulent flows depending on the most important dimensionless parameter which characterizes the particle inertia, namely, the Stokes number (in averaged motion and in large-scale and small-scale fluctuation motion). Note that, by analogy with other parameters fluctuating in turbulent flow, the instantaneous (actual) value of the Reynolds number of a particle $\tilde{R}e_{\mathrm{p}}$ is taken to consist of averaged (time constant) Reynolds number Re_{p} and its fluctuation (time varible) component Re'_{p} : $\tilde{R}e_{\mathrm{p}}(\tau) = Re_{\mathrm{p}} + Re'_{\mathrm{p}}$.

Note that the sign of "=" in Table 1.1 is fairly conventional, because it is clear that inertial particles cannot fully trace either the averaged or, the more so, fluctuation motion of gas. Therefore, we can assume, for correctness, that the particle which fully traces the averaged (fluctuation) flow of gas is a particle whose averaged and fluctuation velocities differ by not more than 1% from the respective velocities of the carrier phase. A similar assumption should be made in treating the processes of heat transfer.

The suggested classification of turbulent heterogeneous flows is quite universal. First, it covers the entire range of particle concentrations from the case of motion of single particles, when their presence has no effect on the characteristics of flow of carrier gas, to the motion of dense sets of particles, when the space taken up by the dispersed phase is comparable to the volume taken up by gas. Second, the classification covers the entire range of particle inertia from minute particles, whose size is commensurable with that of carrier gas molecules, to large stationary particles. In view of the foregoing, this classification of particles cannot be expanded but can only be refined.

Note that, by virtue of the specific features of such flows referred to in the foreword, the theoretical approaches and physical and mathematical models

Table 1.1. Classification of turbulent heterogenous flows by values of the Stokes number (numerals indicate the following forms of flow: 1 – equilibrium flow, 2 – quasiequilibrium flow, 3 –nonequilibrium flow, 4 – flow with large particles, 5– flow past a stationary "frozen" particles)

type of flow	Stokes number	Momentum transfer		characteristic time	heat Transfer
		Reynolds number	particle velocity		particle temperature
1	$Stk_f \to 0$ $Stk_L \to 0$ $Stk_K \approx O(1)$	$Re_p = 0$ $(\overline{Re_p'^2})^{1/2} = 0$	$V = U$ $(\overline{v'^2})^{1/2} = (\overline{u'^2})^{1/2}$	$\tau_p \to 0$	$T_p = T$ $(\overline{t_p'^2})^{1/2} = (\overline{t'^2})^{1/2}$
2	$Stk_f \to 0$ $Stk_L \approx O(1)$ $Stk_K \approx O(1)$	$Re_p = 0$ $(\overline{Re_p'^2})^{1/2} \neq 0$	$V = U$ $(\overline{v'^2})^{1/2} \neq (\overline{u'^2})^{1/2}$	$\tau_p/T_f \to 0$	$T_p = T$ $(\overline{t_p'^2})^{1/2} \neq (\overline{t'^2})^{1/2}$
3	$Stk_f \approx O(1)$ $Stk_L \approx O(1)$ $Stk_K \to \infty$	$Re_p \neq 0$ $(\overline{Re_p'^2})^{1/2} \neq 0$	$V \neq U$ $(\overline{v'^2})^{1/2} \neq (\overline{u'^2})^{1/2}$	$\tau_p \approx O(T_L)$ $\tau_p \approx O(T_f)$	$T_p \neq T$ $(\overline{t_p'^2})^{1/2} \neq (\overline{t'^2})^{1/2}$
4	$Stk_f \approx O(1)$ $Stk_L \to \infty$ $Stk_K \to \infty$	$Re_p \neq 0$ $(\overline{Re_p'^2})^{1/2} = (\overline{u'^2})^{1/2} d_p/\nu$	$V \neq U$ $(\overline{v'^2})^{1/2} = 0$	$\tau_p/T_L \to \infty$	$T_p \neq T$ $(\overline{t_p'^2})^{1/2} = 0$
5	$Stk_f \to \infty$ $Stk_L \to \infty$ $Stk_K \to \infty$	$Re_p = U d_p/\nu$ $(\overline{Re_p'^2})^{1/2} = (\overline{u'^2})^{1/2} d_p/\nu$	$V = U$ $(\overline{v'^2})^{1/2} = 0$	$\tau_P \to \infty$	$T_p = 0$ $(\overline{t_p'^2})^{1/2} = 0$

developed in the mechanics of heterogeneous media may apparently be employed only in a certain, rather narrow, range of particle concentrations and inertia. Because of this, the developed classification of heterogeneous flows is of great methodological and prognostic importance, because it enables one to determine the "cells," i.e., classes of flows, for which the currently available approaches may be acceptable, as well as those for which such approaches are still to be developed. This will be shown in the second chapter.

2

Mathematical Simulation
of Particle-Laden Gas Flows

2.1 Preliminary Remarks

In studying the processes of motion of disperse impurity in the form of solid particles and its inverse effect on the characteristics of turbulence of the carrier continuum, an important part is played by methods of mathematical simulation. Numerous modes of flow of gas suspension, an attempt at classifying which is described in Sect. 1.5, served a basis for the development of a large number of mathematical models of such flows. In constructing models of heterogeneous flows of the most diverse classes, investigators always face an alternative. On the one hand, it is necessary to take into account as many as possible physical processes occurring in heterogeneous flows, which often brings about an undue complication of mathematical formalization of the phenomena being treated. On the other hand, the detailing of a large number of processes the information about each one of these processes is not always indisputable may result in a lower reliability of the model being developed.

It is the objective of this chapter to describe the presently available methods of mathematical simulation of heterogeneous flows. The models of heterogeneous flows of the main types and the characteristic features of simulation of turbulent particle-laden flows of different classes are treated in Sect. 2.2. Section 2.3 is devoted to the description of the possibilities of studying the behavior of solid particles in a turbulent gas flow using two different approaches, namely, stochastic Lagrangian approach and Eulerian continuum approach. The characteristic features of mathematical simulation of gas flow in view of the inverse effect of particles on the flow characteristics are treated in Sect. 2.4.

2.2 Special Features of Simulation of Heterogeneous Flows of Different Types

The wide range of presently existing mathematical models of heterogeneous flows may be divided into two major classes (types). The models of the first class describe the motion of the carrier gas phase and the motion of a plurality of suspended particles and are based on the Eulerian continuum approach. The models of the other type are those based on the Eulerian–Lagrangian description of motion of heterogeneous medium, namely, the equations of motion of the gas phase are solved in the Eulerian formulation, while the motion of particles is described by Lagrangian equations which are integrated along their trajectories.

It is clear that attempts at making an adequate description of the entire diversity of heterogeneous flows using models of both types mentioned earlier are hardly justified. Therefore, for certain classes of flows (see Sect. 1.5), which are first of all characterized by the concentration of disperse impurity and its inertia (Stokes number), models of one or the other type must be preferred.

We will consider briefly the advantages and limitations of the Eulerian (two-fluid) and Eulerian–Lagrangian models of description of gas–solid flows [28, 52, 58].

The advantage of two-fluid models is the use of like equations for the description of the gas and dispersed phases. This enables one to utilize the rich experience of simulation of single-phase turbulent flows and apply the same numerical methods of solving the entire set of equations. The disadvantages of such models include some "loss" of information about the motion of individual particles, as well as the difficulties in the formulation of boundary conditions for the dispersed phase on surfaces which bound the flow.

We will now turn to the Eulerian–Lagrangian models. The advantage of these models consists in the possibility of obtaining detailed statistical information about the motion of individual particles as a result of integration of equations of motion (heat transfer) of particles in a known (pre-calculated) velocity (temperature) field of carrier gas. However, as the concentration of the dispersed phase increases, difficulties arise which are associated with the use of the Eulerian–Lagrangian models. Two aspects may be identified in this respect. First, the concentration increase leads to the inverse effect of particles on the carrier gas parameters, and the calculations need to be performed in several iterations; as a result, the computation procedure is complicated. Second, the concentration increase causes a rise of the probability of particle collisions with one another, which brings about entanglement of their trajectories. As the particle size decreases, the use of trajectory methods for the calculation of particle motion is also complicated. This is associated with the fact that it is necessary to take into account the interaction between the particles and turbulent eddies of ever smaller dimensions in order to obtain correct information about the averaged characteristics of the dispersed phase. The latter fact further complicates the computations.

Flows of two extreme classes exist (see Sect. 1.5), namely, flows with particles of extremely low inertia (the case of equilibrium flow) and flows with an extremely low concentration of the dispersed phase (the mode with single particles, in which their presence has no effect on the carrier gas flow). Simplified mathematical models may be employed for flows of these classes, namely, a one-velocity one-temperature diffusion model (Eulerian approach) for low-inertia particles and a single-particle approximation (Lagrangian approach) for a low-concentration flow.

In the case of increasing concentration and inertia of particles, it is not a simple problem to choose between two types of models of heterogeneous flows. Therefore, the types of heterogeneous flows which are most complex from the standpoint of mathematical simulation are flows of "intermediate" classes. According to the classification given earlier (see Sect. 1.5), such flows are nonequilibrium flows and flows with large particles at moderate values of volume concentration of the dispersed phase, when the presence of particles affects all (without exception) characteristics of carrier gas.

Treating the hydrodynamics of flows of a special class such as the flow past a stationary "frozen" particle (see Sect. 1.5), a peculiar analog of which is the flow of a single-phase liquid (gas) past tube bundles, falls outside of the scope of this monograph.

When one tries to use two-fluid models, the question arises first of all whether it is possible to use the methods of continuum dynamics to describe the motion of a plurality of particles [39]. A continual description for an ensemble of particles is possible in the case where a geometric scale may be indicated which, on the one hand, is negligible compared to the scale of variation of the flow parameters and, on the other hand, is large enough to contain a significant number of particles which permits a correct determination of their averaged parameters [10]. We will make the simplest estimates which enable one to determine such a geometric scale for a heterogeneous flow with particles of diameter d_p and volume concentration Φ. For this purpose, we will treat an element of flow in the form of a cube with edge a, which contains N_p particles. The expression for the volume concentration of particles will be written as:

$$\Phi = \frac{\pi d_p^3 N_p}{6a^3}. \tag{2.1}$$

We use (2.1) to find the formula for the ratio of the cube edge to the particle diameter,

$$\frac{a}{d_p} = \sqrt[3]{\frac{\pi N_p}{6\Phi}}. \tag{2.2}$$

The dependence of the relative dimension of the edge of a cube containing N_p particles on the value of the volume concentration of the dispersed phase, obtained by relation (2.2), is given in Fig. 2.1. The calculations were performed for two values of the number of particles in the flow volume of interest to us,

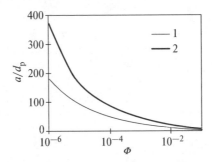

Fig. 2.1. The relative dimension of the edge of a cube as a function of the value of the volume concentration of particles: (1) $N_p = 10$, (2) $N_p = 100$

$N_p = 10$ and 100. Obviously, the relative fluctuation of distributed density of the dispersed phase in the volume being treated increases with decreasing number of particles and reaches several percent at $N_p = 100$. If the foregoing error in determining the particle density is inadequate, a plurality of particles cannot be regarded as a continuum on scales comparable to a or lower. In this case, the motion of particles cannot be described by the methods of continuum dynamics.

The data given in Fig. 2.1 indicate that the scale a increases with decreasing volume concentration of particles and with increasing particle size. For example, for particles $50\,\mu\mathrm{m}$ in diameter ($\Phi = 10^{-3}$), the scale is $a \approx 1.9\,\mathrm{mm}$, and for particles $100\,\mu\mathrm{m}$ in diameter ($\Phi = 10^{-4}$) – $a \approx 8\,\mathrm{mm}$.

Therefore, the general tendency is as follows: as the concentration of particles increases and their inertia decreases, the Eulerian continuum approach turns out to be preferable for use in describing the dynamics of disperse impurity.

2.3 Description of Motion of Solid Particles Suspended in Turbulent Flow

The motion of particles suspended in a turbulent gas flow may be calculated both within the frame of stochastic Lagrangian approach and using Eulerian continuum approach.

2.3.1 Lagrangian Approach

The study of regularities of the behavior of particles in the known velocity field of the carrier phase is of interest per se when calculating weakly dusty flows without the inverse effect of the dispersed phase on the characteristics of gas and may also be an integral part of the process of construction of complex mathematical models for the description of heterogeneous flows of most diverse classes.

The Lagrangian equation of instantaneous motion of a single solid particle in a turbulent gas flow has the form:

$$\rho_{\mathrm{p}}\frac{\pi d_{\mathrm{p}}^3}{6}\frac{\mathrm{d}v_i}{\mathrm{d}\tau} = \sum_i f_i(r_{\mathrm{p}},\tau),\qquad(2.3)$$

where $f_i(r_{\mathrm{p}},\tau)$ denotes the external forces acting on the particle, and r_{p} is the particle coordinate.

The main force factors affecting the motion of the dispersed phase will be treated later.

Aerodynamic Drag Force

This force arises due to the difference between the velocity of gas and the velocity of a particle moving in this gas (see Fig. 2.2). The effect of the aerodynamic drag force causes the particle acceleration if $U > V$ and, on the contrary, the deceleration in the case of $U < V$. The expression for aerodynamic force has the form:

$$\overrightarrow{F}_{\mathrm{A}} = C_{\mathrm{D}}\rho\frac{\pi d_{\mathrm{p}}^2}{4}\frac{|\overrightarrow{U} - \overrightarrow{V}|(\overrightarrow{U} - \overrightarrow{V})}{2},\qquad(2.4)$$

where the particle drag coefficient in the case of incompressible flow is a function of the Reynolds number, i.e., $C_{\mathrm{D}} = C_{\mathrm{D}}(Re_{\mathrm{p}})$. The graph of this dependence is often referred to as standard drag curve. Numerous formulas are available in the literature, which approximate this curve for different ranges of the Reynolds number [39,46]. For low values of the Reynolds number ($Re_{\mathrm{p}} < 1$), the well-known Stokes formula is valid,

$$C_{\mathrm{D}} = \frac{24}{Re_{\mathrm{p}}},\quad Re_{\mathrm{p}} = \frac{|\overrightarrow{U} - \overrightarrow{V}|d_{\mathrm{p}}}{\nu}.\qquad(2.5)$$

The equation of averaged motion of a Stokesian particle has the form

$$\frac{\mathrm{d}V_i}{\mathrm{d}\tau} = \frac{U_i - V_i}{\tau_{\mathrm{p}0}}\qquad(2.6)$$

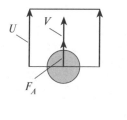 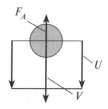

Fig. 2.2. A scheme of particle motion under the effect of the aerodynamic drag force

where τ_{p0} is the time of dynamic relaxation of the Stokesian particle (see Sect. 1.4).

As the Reynolds number increases ($Re_p \geq 1$), the value of the particle drag coefficient deviates from the Stokes law toward higher values, while the particle relaxation time, on the contrary, decreases. For taking this fact into account, the correction function $C = C(Re_p)$ is introduced. The values of this function are given in Sect. 1.4. Expression (2.6) takes the following form for a non-Stokesian particle:

$$\frac{dV_i}{d\tau} = \frac{U_i - V_i}{\tau_p}, \tag{2.7}$$

where $\tau_p = \tau_{p0}/C$.

Equation (2.7) of averaged motion of a non-Stokesian particle is very approximate, because it does not include the effect of turbulent fluctuations of the carrier phase.

Note that the standard curve describes the drag of single smooth spherical particles during their uniform motion in a laminar flow of liquid (gas). The problems associated with the inclusion of the effect made on the drag of the dispersed phase by the asphericity of particles, by the state of their surface, by the degree of flow turbulence, by the concentration and geometric constraint of motion, and by other factors, were treated in [39, 46].

Gravity Force

Along with the aerodynamic drag force, this force is one of the most important force factors defining the dynamics of particles. The expression for gravity force has the form:

$$\overrightarrow{F}_g = \rho_p \frac{\pi d_p^3}{6} \overrightarrow{g}. \tag{2.8}$$

The effect of gravity force on particle motion will be significant, and its inclusion is necessary in the case where the free-fall velocity of particles and the velocity of flow in which they are suspended are quantities of the same order of magnitude.

Saffman Force

This force arises because of the nonuniformity of the profile of averaged velocity of carrier gas. The difference between the relative velocities of flow past a particle on different sides results in the emergence of a pressure difference. The particle will move toward lower pressure (see Fig. 2.3). The value of the Saffman force acting on a particle during its motion in a laminar flow with a linear velocity profile is determined as follows [38]:

$$F_S = k_S \nu^{1/2} \rho d_p^2 (U_x - V_x) \left(\frac{dU_x}{dr}\right)^{1/2}. \tag{2.9}$$

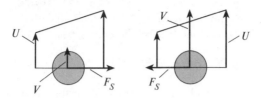

Fig. 2.3. A scheme of transverse migration of a particle in a nonuniform flow under the effect of the Saffman force

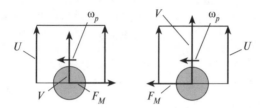

Fig. 2.4. A scheme of migration of a rotating particle under the effect of the Magnus force

In the case of $U_x/(\nu\, dU_x/dr)^{1/2} \ll 1$, the value of the coefficient in (2.9) is $k_S = 1.61$.

The Saffman force may have a significant effect on the particle motion in the wall region where high gradients of averaged velocity of carrier gas are observed.

Magnus Force

Its emergence is due to the particle rotation. During their motion in a gas flow, particles of complex shape (aspherical) always rotate. As to spherical particles, they will also rotate in a flow with a nonuniform velocity profile. A rotating particle entrains the gas. As a result, the pressure on the side where the directions of flow past the particle and rotation of gas elements coincide becomes lower compared to the region in which these directions are opposite. Therefore, the particle will move toward lower pressure (see Fig. 2.4). The magnitude of the force acting on a particle during its rotation in a laminar flow with a uniform velocity profile at $Re_p = |\overrightarrow{W}|d_p/\nu \ll 1$ and $Re_\omega = |\overrightarrow{\omega}_p|d_p^2/\nu \ll 1$ is defined by the following expression [37]:

$$\overrightarrow{F}_M = k_M \rho \left(\frac{d_p}{2}\right)^3 (\overrightarrow{W} \times \overrightarrow{\omega}_p). \qquad (2.10)$$

Here, ω_p is the rotational velocity of the particle. For the foregoing values of the Reynolds number, the coefficient in (2.10) is $k_M = \pi$. For the other limiting case of high values of the Reynolds number ($Re_p \to \infty, Re_\omega \to \infty$), this coefficient becomes $k_M = 8\pi/3$ [34].

For the range of moderate values of the Reynolds number, the following expression may be recommended for the calculation of the coefficient [55]:

$$k_{\mathrm{M}} = 0.534 Re_{\omega}^{-0.64} Re_{\mathrm{p}}^{0.715}. \tag{2.11}$$

The use of relation (2.11) enables one to describe the majority of available calculation and experimental data in the Reynolds number range of $590 < Re_{\omega} < 45,000$ and $360 < Re_{\mathrm{p}} < 13,500$.

Shraiber et al. [39] analyzed the effect of the Magnus force on the particle motion. They showed the Magnus force to be almost always less than the Saffman force. Nevertheless, it is wrong to ignore the transverse shift of particles due to the effect of the Magnus force in high-velocity flows in which high gradients of gas velocity are realized and, consequently, high rotational velocities of particles.

Turbophoresis Force

This force arises because of the nonuniformity of the profile of fluctuation velocity of carrier gas. The gradient of the profile of the transverse component of fluctuation velocity of gas (see Sect. 1.3) leads to a directional shift of a particle toward decreasing intensity of fluctuations (see Fig. 2.5). The expression for the turbophoresis force acting on a particle has the form [31]

$$F_{\mathrm{Tu}} = -\frac{1}{2}\rho_{\mathrm{p}}\frac{\pi d_{\mathrm{p}}^{3}}{6}\frac{\partial \overline{u_r'^2}}{\partial r}. \tag{2.12}$$

This force may bring about a significant transverse displacement of a particle during its motion in the wall region.

Thermophoresis Force

This force arises as a result of the nonuniformity of the temperature profile of carrier gas. The gas molecules make a more intense force effect on a particle on its higher-temperature side. Therefore, the particle tends to move from

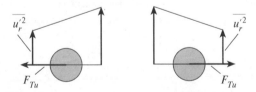

Fig. 2.5. A scheme of displacement of a particle in a nonuniform field of fluctuation velocity of gas under the effect of the turbophoresis force

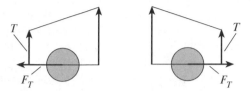

Fig. 2.6. A scheme of motion of a particle in a nonuniform temperature field under the effect of the thermophoresis force

the more heated to less heated regions (see Fig. 2.6). The expression for the thermophoresis force acting on a particle of low thermal conductivity has the form [19]:

$$F_{\mathrm{T}} = -\frac{4.5\rho\nu^2 d_{\mathrm{p}}\lambda}{T(2\lambda + \lambda_{\mathrm{p}})}\frac{\partial T}{\partial r}. \tag{2.13}$$

More theoretical formulas have been suggested for determining the value of thermophoretic force. The most complete inventory of the available relations is found in [31, 43].

Note a very important point. One must know the instantaneous values of forces in order to calculate the actual velocity of particles in accordance with (2.3). The foregoing formulas make it possible to determine only some averaged values of the force factors acting on the particles, because they fully ignore the turbulent fluctuations of gas velocity (temperature). The question of the effect of turbulence of the dispersed phase on magnitude of the forces remains open.

Shraiber et al. [39] and Gavin and Shraiber [22] tried to determine the fluctuation values of forces by applying the Reynolds procedure and using the thus derived expressions to construct equations of fluctuation motion and heat transfer of particles. However, the expressions obtained for the averaged and fluctuation values of forces are, in my opinion, too cumbersome and cannot be recommended for use.

Lagrangian Equations of Fluctuation Motion and Heat Transfer for Particles

For the case where the main effect on the particle motion is made by the aerodynamic drag and gravity forces, the Lagrangian equations of motion and heat transfer have the form

$$\frac{\mathrm{d}v_i}{\mathrm{d}\tau} = \frac{u_i - v_i}{\tau_{\mathrm{p}}} \pm g, \tag{2.14}$$

$$\frac{\mathrm{d}t_{\mathrm{p}}}{\mathrm{d}\tau} = \frac{t - t_{\mathrm{p}}}{\tau_{\mathrm{t}}}. \tag{2.15}$$

We will derive equations of fluctuation motion and heat transfer for inertial particles. The difficulties associated with the construction of such equations for the case of nonlinear law of aerodynamic drag were treated in detail by Shraiber et al. [39]. The developed approach to the derivation of fluctuation equations for the dispersed phase is based on the application of the Reynolds procedure to actual Lagrangian equations for particles. The results given later were borrowed from [48, 51], where the method described earlier was used to derive and analyze the approximate one-dimensional equations for fluctuations of velocity and temperature of the dispersed phase during the realization of heterogeneous flows of different classes.

We will make the following assumptions for analysis (1) the case of weakly dusty flows is treated, where the particles have little effect on one another; (2) the particles have a spherical shape; (3) the particle motion is defined by the effect of only two force factors, namely, the aerodynamic drag and gravity forces; (4) the fluctuations of the physical properties of carrier gas are ignored; (5) assumption is made of the additivity of the averaged and fluctuation dynamic slip between the phases in determining the instantaneous value of the particle drag coefficient; (6) the heat transfer between the particles and the carrier phase is defined by the convection component alone; and (7) the temperature gradient within a particle is negligible.

We will rewrite the equations of one-dimensional motion and heat transfer of a particle (2.14) and (2.15) in instantaneous (actual) variables as:

$$\frac{dv_x}{d\tau} = \frac{u_x - v_x}{\tau_p} \pm g, \tag{2.16}$$

$$\frac{dt_p}{d\tau} = \frac{t - t_p}{\tau_t}, \tag{2.17}$$

where

$$\tau_p = \frac{\tau_{p0}}{C} = \frac{\rho_p d_p^2}{18\mu C}, \quad C = 1 + \frac{1}{6}\tilde{Re}_p^{2/3},$$

$$\tilde{Re}_p = \frac{|u_x - v_x|d_p}{\nu}, \quad \tau_t = \frac{\tau_{t0}}{C_1} = \frac{C_p \rho_p d_p^2}{12\lambda C_1},$$

$$C_1 = 1 + 0.3\,\tilde{Re}_p^{1/2} Pr^{1/3}, \tilde{Re}_p \le 10^3.$$

We will represent the actual velocities and temperatures of the particle and carrier gas in the form of sums of respective averaged and fluctuation components,

$$v_x = V_x + v'_x, \tag{2.18}$$

$$u_x = U_x + u'_x, \tag{2.19}$$

$$t_p = T_p + t'_p, \tag{2.20}$$

$$t = T + t'. \tag{2.21}$$

We will treat the instantaneous Reynolds number of the particle similarly,

$$\tilde{Re}_\mathrm{p} = Re_\mathrm{p} + Re'_\mathrm{p}, \qquad (2.22)$$

where $Re_\mathrm{p} = \frac{|U_x - V_x| d_\mathrm{p}}{\nu}$ and $Re'_\mathrm{p} = \frac{|u'_x - v'_x| d_\mathrm{p}}{\nu}$.

We will substitute (2.18)–(2.22) into (2.16) and (2.17) and perform the averaging procedure on the resultant equations. The equations of averaged motion and heat transfer of the dispersed phase will take the form:

$$\frac{\mathrm{d}V_x}{\mathrm{d}\tau} = \frac{U_x - V_x}{\tau_{\mathrm{p}0}} + \frac{1}{6\tau_{\mathrm{p}0}} \left[\overline{(U_x - V_x)(Re_\mathrm{p} + Re'_\mathrm{p})^{2/3}} \right.$$
$$\left. + \overline{(u'_x - v'_x)(Re_\mathrm{p} + Re'_\mathrm{p})^{2/3}} \right] \pm g, \qquad (2.23)$$

$$\frac{\mathrm{d}T_\mathrm{p}}{\mathrm{d}\tau} = \frac{T - T_\mathrm{p}}{\tau_{\mathrm{t}0}} + \frac{0.3 Pr^{1/3}}{\tau_{\mathrm{t}0}} \left[\overline{(T - T_\mathrm{p})(Re_\mathrm{p} + Re'_\mathrm{p})^{1/2}} \right.$$
$$\left. + \overline{(t' - t'_\mathrm{p})(Re_\mathrm{p} + Re'_\mathrm{p})^{1/2}} \right]. \qquad (2.24)$$

We will subtract (2.23) and (2.24) term-by-term from (2.16) and (2.17), respectively, in view of the substitution of (2.18)–(2.22) into the latter equations, to derive equations of fluctuation motion and fluctuation heat transfer for particles,

$$\frac{\mathrm{d}v'_x}{\mathrm{d}\tau} = \frac{u'_x - v'_x}{\tau_{\mathrm{p}0}} + \frac{U_x - V_x}{6\tau_{\mathrm{p}0}} \left[(Re_\mathrm{p} + Re'_\mathrm{p})^{2/3} - \overline{(Re_\mathrm{p} + Re'_\mathrm{p})^{2/3}} \right]$$
$$+ \frac{1}{6\tau_{\mathrm{p}0}} \left[(u'_x - v'_x)(Re_\mathrm{p} + Re'_\mathrm{p})^{2/3} - \overline{(u'_x - v'_x)(Re_\mathrm{p} + Re'_\mathrm{p})^{2/3}} \right],$$
$$(2.25)$$

$$\frac{\mathrm{d}t'_\mathrm{p}}{\mathrm{d}\tau} = \frac{t' - t'_\mathrm{p}}{\tau_{\mathrm{t}0}} + \frac{0.3 Pr^{1/3}}{\tau_{\mathrm{t}0}} \left\{ (T - T_\mathrm{p}) \left[(Re_\mathrm{p} + Re'_\mathrm{p})^{1/2} - \overline{(Re_\mathrm{p} + Re'_\mathrm{p})^{1/2}} \right] \right.$$
$$\left. + \left[(t' - t'_\mathrm{p})(Re_\mathrm{p} + Re'_\mathrm{p})^{1/2} - \overline{(t' - t'_\mathrm{p})(Re_\mathrm{p} + Re'_\mathrm{p})^{1/2}} \right] \right\}. \qquad (2.26)$$

It is difficult to use the resultant equations of fluctuation motion and heat transfer for particles (2.25) and (2.26), as well as the respective averaged equations (2.23) and (2.24), for calculations by virtue of indeterminacy of the correlation terms. In [48, 51], (2.25) and (2.26) for particle-laden flows of different classes were analyzed (see Sect. 1.5). The results obtained in [48, 51] will be given later.

Quasiequilibrium flow. We will treat two possible versions of realization of quasiequilibrium flow. The first version involves a flow with a low fluctuation slip of particles ($Re'_\mathrm{p} < 1$). In this case, the drag of particles obeys the Stokes law. The second version involves a flow with a relatively high slip of the dispersed phase in fluctuation motion ($1 \leq Re'_\mathrm{p} < 1{,}000$). For this case, the correction to the Stokes law of resistance must be taken into account.

In view of the fact that, in the case of quasiequilibrium flow, the averaged dynamic and thermal slip is zero ($Re_p = 0, T - T_p = 0$), (2.25) and (2.26) may yield, for the case of Stokesian particles [48, 51],

$$\frac{dv'_x}{d\tau} = \frac{u'_x - v'_x}{\tau_{p0}}, \tag{2.27}$$

$$\frac{dt'_p}{d\tau} = \frac{t' - t'_p}{\tau_{t0}}. \tag{2.28}$$

Approximate equations of fluctuation motion and heat transfer for particles for the case where the fluctuation slip is significant have the form [48, 51]

$$\frac{dv'_x}{d\tau} = \frac{u'_x - v'_x}{\tau_{p0}} \left(1 + \frac{1}{6} Re_p'^{2/3} \right), \tag{2.29}$$

$$\frac{dt'_p}{d\tau} = \frac{t' - t'_p}{\tau_{t0}} \left(1 + 0.3 Re_p'^{2/3} Pr^{1/3} \right). \tag{2.30}$$

Nonequilibrium flow. In this case, it does not appear possible to ignore the interphase slip in averaged or fluctuation motion, because the values of slip in these motions often turn out to be of the same order of magnitude, i.e., $O(Re_p'/Re_p) = 1$.

In view of assumptions made in [48, 51] and in order to simplify analysis of the correlation terms, the approximate equations of fluctuation motion and heat transfer (2.25) and (2.26) for nonequilibrium flow take the form:

$$\frac{dv'_x}{d\tau} = \frac{u'_x - v'_x}{\tau_{p0}} \left[1 + \frac{1}{6} (Re_p + Re_p')^{2/3} \right], \tag{2.31}$$

$$\frac{dt'_p}{d\tau} = \frac{t' - t'_p}{\tau_{t0}} \left[1 + 0.3 (Re_p + Re_p')^{1/2} Pr^{1/3} \right]. \tag{2.32}$$

It follows from (2.31) and (2.32) that the averaged slip causes an increase in the fluctuation velocity and temperature of particles.

Flow with large particles. Under conditions of this flow, the averaged slip between the phases is far beyond the fluctuation slip, i.e., $Re_p'/Re_p \to 0$. In this case, the inertia of particles is so high that they hardly take part either in fluctuation motion ($v'_x = 0$) or in fluctuation heat transfer ($t'_p = 0$). The following trivial notation of (2.25) and (2.26) was obtained in [48, 51], using some assumptions, for flows of this class:

$$\frac{dv'_x}{d\tau} = 0, \tag{2.33}$$

$$\frac{dt'_p}{d\tau} = 0. \tag{2.34}$$

The foregoing approximate equations of fluctuation motion and fluctuation heat transfer for particles are of interest per se and may be used to determine the fluctuation velocity and temperature of particles. For this purpose, the resultant equations are integrated with respect to time. This time is the minimal of three times [49, 54, 56], namely, (1) the time of dynamic (thermal) relaxation of particles, (2) the time of interaction between particles and energy-carrying turbulent eddies of carrier gas, and (3) the lifetime of turbulent eddy.

At first glance it would seem that the obtained relations may also be employed to construct equations for correlations associated with the dispersed phase. Such correlations are present in equations which describe the carrier gas motion (see Sect. 2.4). It is necessary to calculate these correlations for assessing the inverse effect of particles on the parameters of gas flow. However, equations of motion of the carrier medium are written using the Eulerian continuum approach. Consequently, the correlations appearing in these equations must also be derived using Euler's method [39]. As to the method of constructing equations of fluctuation motion and heat transfer for particles, which is described earlier, it is purely Lagrangian; therefore, the resultant equations cannot be used to study the inverse effect of particles within the Eulerian approach.

The possibilities of using the Lagrangian trajectory method for studying the behavior of particles in turbulent gas flows may be well illustrated by studies [40–42].

2.3.2 Eulerian Continuum Approach

We will now consider the presently existing approaches to the construction of continuum equations of particle motion and analyze the singularities of the description of behavior of the dispersed phase for heterogeneous flows of different classes.

Equations describing the averaged motion and heat transfer of particles are written by analogy with the equations for gas (1.6)–(1.8). The set of equations for the dispersed phase also turns out to be nonclosed, because the equations contain second moments for the fluctuations of velocity $\overline{v_i' v_j'}$, as well as of velocity and temperature $\overline{v_j' t_{\mathrm{p}}'}$ of particles, similar to Reynolds stresses and turbulent heat flux in gas. Based on the experience of studying single-phase flows, various models are used for closing the set of averaged equations of motion and heat transfer for particles. The best known are the algebraic and differential models.

Two basic approaches exist to determining the correlations of velocity of the dispersed phase within the algebraic models. According to the first approach, the correlation moments are expressed directly in terms of Reynolds stresses of the carrier flow [23, 26],

$$\overline{v_i' v_j'} = A \overline{u_i' u_j'}, \tag{2.35}$$

where A is the function of involvement of particles in the fluctuation motion of gas.

Expression (2.35) is valid for relatively small particles (quasiequilibrium flow) under conditions of uniform distribution of the averaged velocity of the dispersed phase in the flow.

The second method of determining turbulent stresses in the dispersed phase is by using gradient relations of the Boussinesq type for single-phase flow [11],

$$\overline{v_i' v_j'} = -\nu_{\mathrm{p}} \left(\frac{\partial V_i}{\partial x_j} \right), \tag{2.36}$$

or in the form [32, 52]

$$\overline{v_i' v_j'} = -\nu_{\mathrm{p}} \left(\frac{\partial V_i}{\partial x_j} + \frac{\partial V_j}{\partial x_i} - \frac{2}{3} \frac{\partial V_k}{\partial x_k} \delta_{ij} \right) + \frac{2}{3} k_{\mathrm{p}} \delta_{ij}, \tag{2.37}$$

where ν_{p} is the coefficient of turbulent viscosity of the dispersed phase. Various methods of determining ν_{p} are described in the literature [32, 52].

Along with the algebraic models, the differential models are extensively employed at present to describe the turbulent momentum and heat transfer in the dispersed phase. These models are based on the use of equations of energy balance of fluctuations of the dispersed phase or of the second moments of fluctuations of particle velocity and temperature.

A consistent method of constructing Euler's equations of motion and heat transfer for the dispersed phase in a turbulent flow is the method based on the use of a kinetic equation for the probability density function (PDF) of particle velocity and temperature [12, 13, 35, 57]. According to this approach, the probability density of particle distribution by coordinates $\overrightarrow{x}$, velocities $\overrightarrow{v}$, and temperatures t_{p} is introduced for making a transition from stochastic equations of the Langevin type (such as equations of instantaneous motion and heat transfer for a single particle) to a kinetic equation for a plurality of particles,

$$P(\overrightarrow{x}, \overrightarrow{v}, t_{\mathrm{p}}, \tau) = \overline{\delta(\overrightarrow{x} - \overrightarrow{r}_{\mathrm{p}}(\tau))\delta(\overrightarrow{v} - \overrightarrow{v}_{\mathrm{p}}(\tau))\delta(t - t_{\mathrm{p}}(\tau))}, \tag{2.38}$$

where averaging is performed over realizations of random fields of velocity and temperature of carrier gas. Then, the differentiation of (2.38) with respect to time in view of representation of the gas velocity and temperature in the instantaneous equations of motion and heat transfer for particles in the form of sums of averaged and fluctuation components is used to derive the equation for probability density. Then, the equation for the PDF of particle distribution by coordinates, velocities, and temperatures is used to construct equations for averaged concentration, velocity, and temperature of particles, which have the form [52]:

$$\frac{\partial \Phi}{\partial \tau} + \sum_j \frac{\partial \Phi V_j}{\partial x_j} = 0, \tag{2.39}$$

$$\frac{\partial V_i}{\partial \tau} + \sum_j V_j \frac{\partial V_i}{\partial x_j} = -\sum_j \frac{\partial \overline{v_i' v_j'}}{\partial x_j} + \frac{U_i - V_i}{\tau_{\mathrm{p}}} - \sum_j \frac{D_{\mathrm{p}ij}}{\tau_{\mathrm{p}}} \frac{\partial \ln \Phi}{\partial x_j}, \quad (2.40)$$

$$\frac{\partial T_{\mathrm{p}}}{\partial \tau} + \sum_j V_j \frac{\partial T_{\mathrm{p}}}{\partial x_j} = -\sum_j \frac{\partial \overline{v_j' t_{\mathrm{p}}'}}{\partial x_j} + \frac{T - T_{\mathrm{p}}}{\tau_{\mathrm{t}}} - \sum_j \frac{D_{\mathrm{p}j}^{\mathrm{t}}}{\tau_{\mathrm{t}}} \frac{\partial \ln \Phi}{\partial x_j}, \quad (2.41)$$

where

$$\overline{v_i' v_j'} = \frac{1}{\Phi} \iint v_i' v_j' P \mathrm{d}v \mathrm{d}t_{\mathrm{p}}, \quad \overline{v_j' t_{\mathrm{p}}'} = \frac{1}{\Phi} \iint v_j' t_{\mathrm{p}}' P \mathrm{d}v \mathrm{d}t_{\mathrm{p}},$$

$$D_{\mathrm{p}ij} = \tau_{\mathrm{p}}(\overline{v_i' v_j'} + g_{\mathrm{p}} \overline{u_i' u_j'}), \quad D_{\mathrm{p}j}^{\mathrm{t}} = \tau_{\mathrm{t}} \overline{v_j' t_{\mathrm{p}}'} + \tau_{\mathrm{p}} g_{\mathrm{pt}} \overline{u_j' t'},$$

$$g_{\mathrm{p}} = \frac{T_{\mathrm{pL}}}{\tau_{\mathrm{p}}} - 1 + \exp(-T_{\mathrm{pL}}/\tau_{\mathrm{p}}),$$

$$g_{\mathrm{pt}} = \frac{T_{\mathrm{pLt}}}{\tau_{\mathrm{p}}} - 1 + \exp(-T_{\mathrm{pLt}}/\tau_{\mathrm{p}}).$$

Here, T_{pL} and T_{pLt} denote the time of interaction of particles with energy-intensive fluctuations of velocity and temperature, respectively. For an inertialess impurity,

$$T_{\mathrm{pL}} = T_{\mathrm{L}}, \quad T_{\mathrm{pLt}} = T_{\mathrm{Lt}}, \quad (2.42)$$

where T_{L} and T_{Lt} are the time scales of fluctuations of velocity and temperature of gas, respectively.

In the case of nonequilibrium flow, where the averaged and dynamic slips between the gas and particles become significant, the times of interaction with fluctuations of the carrier flow may differ significantly from the respective scales of fluctuations of the carrier phase.

The set of (2.39)–(2.41) is not closed, because the equations include the turbulent stresses $\overline{v_i' v_j'}$ and the turbulent heat flux $\overline{v_j' t_{\mathrm{p}}'}$ in the dispersed phase, associated with the involvement of particles in the fluctuation motion, as well as turbulent diffusion fluxes of momentum and heat arising because of the nonuniformity of the particle concentration.

Volkov et al. [52] developed a mathematical description of the processes of momentum and heat transfer in the dispersed phase of different levels of detail. A closed set of equations is given on the level for the third moments. In this case, the fourth moments of fluctuation characteristics, which appear in equations for the third moments, are expressed approximately in terms of the sum of products of the second moments [52]. Triple correlations must be determined in order to describe the hydrodynamics and heat transfer of the dispersed phase on the level of equations for the second moments. For this purpose, Volkov et al. [52] further used equations for the third moments; the simplification of these equations by ignoring small terms enables one to find algebraic relations for triple correlations which contain only the second moments. The computational scheme may be further simplified by replacing the equations for the second moments of velocity fluctuations by a single differential equation for the energy of fluctuations of the dispersed phase, which has the following form [52]:

$$\frac{\partial k_{\mathrm{p}}}{\partial \tau}+\sum_j V_j \frac{\partial k_{\mathrm{p}}}{\partial x_j}=-\frac{1}{\Phi}\sum_j \frac{\partial \Phi \overline{v_i' v_i' v_j'}}{2\partial x_j}-\sum_j\sum_i \overline{v_i' v_j'}\frac{\partial V_i}{\partial x_j}+\frac{2}{\tau_{\mathrm{p}}}(f_u k-k_{\mathrm{p}}), \quad (2.43)$$

where $k_{\mathrm{p}}=\frac{1}{2}\sum_i \overline{v_i' v_i'}$ is the energy of fluctuations of particle velocity.

In a steady-state uniform flow or for small particles (quasiequilibrium flow), (2.43) yields $k_{\mathrm{p}}=f_u k$, where $f_u=(1+Stk_{\mathrm{L}})^{-1}$. In this case, (2.39) and (2.40) in view of relation (2.37) give a description of momentum transfer in the dispersed phase on the level of equations for the first moments.

2.4 Description of Motion of Gas Carrying Solid Particles

We will treat the motion of gas in the presence of particles when the particles start making an inverse effect on the gas characteristics. The equations of continuity, motion, and energy for the gas phase with a relatively low content of particles ($\varphi \ll 1$) in the absence of external mass forces have the form:

$$\sum_j \frac{\partial u_j}{\partial x_j}=0, \quad (2.44)$$

$$\frac{\partial u_i}{\partial \tau}+\sum_j u_j \frac{\partial u_i}{\partial x_j}=-\frac{1}{\rho}\frac{\partial p}{\partial x_i}+\nu \sum_j \frac{\partial^2 u_i}{\partial x_j \partial x_j}-\frac{\rho_{\mathrm{p}}\varphi\,(u_i-v_i)}{\rho}\,\frac{}{\tau_{\mathrm{p}}}, \quad (2.45)$$

$$\frac{\partial t}{\partial \tau}+\sum_j u_j \frac{\partial t}{\partial x_j}=a\sum_j \frac{\partial^2 t}{\partial x_j \partial x_j}-\frac{C_{\mathrm{p_p}}\rho_{\mathrm{p}}\varphi\,(t-t_{\mathrm{p}})}{C_{\mathrm{p}}\rho}\,\frac{}{\tau_{\mathrm{t}}}. \quad (2.46)$$

The continuity equation (2.44) has a similar form as (1.1) for a single-phase flow. Equations (2.45) and (2.46) differ from the respective equations of motion and energy for a single-phase gas (1.2) and (1.3) by the presence in their right-hand parts of terms which take into account the dynamic and thermal effect of the dispersed phase on the carrier flow.

We will average (2.44)–(2.46) over time. In so doing, we will follow the well-known method of averaging in the theory of single- phase flows of variable density [25], as well as the PDF-based method of constructing equations for the dispersed phase [52], and assume $\overline{\varphi' v_i'}=\overline{\varphi' t_{\mathrm{p}}'}=0$. The averaged equations of continuity, motion, and energy have the form:

$$\sum_j \frac{\partial U_j}{\partial x_j}=0, \quad (2.47)$$

$$\frac{\partial U_i}{\partial \tau} + \sum_j U_j \frac{\partial U_i}{\partial x_j} = -\frac{1}{\rho}\frac{\partial P}{\partial x_i} + \nu \sum_j \frac{\partial^2 U_i}{\partial x_j \partial x_j} - \sum_j \frac{\partial(\overline{u_i'u_j'})}{\partial x_j}$$

$$- \frac{\rho_p \Phi}{\rho}\frac{(U_i - V_i)}{\tau_p} - \frac{\rho_p \overline{\varphi' u_i'}}{\rho \tau_p}, \tag{2.48}$$

$$\frac{\partial T}{\partial \tau} + \sum_j U_j \frac{\partial T}{\partial x_j} = a \sum_j \frac{\partial^2 T}{\partial x_j \partial x_j} - \sum_j \frac{\partial(\overline{u_j't'})}{\partial x_j}$$

$$- \frac{C_{p_p}\rho_p \Phi}{C_p \rho}\frac{(T - T_p)}{\tau_t} - \frac{C_{p_p}\rho_p \overline{\varphi't'}}{C_p \rho \tau_t}. \tag{2.49}$$

Equations (2.48) and (2.49) indicate that the inverse effect of particles on the motion and heat transfer of carrier gas is defined by the averaged dynamic and thermal slip of the dispersed phase, as well as by the fluctuations of the particle concentration. Note that the contribution made by the penultimate and last terms of the right-hand parts of (2.48) and (2.49) will be determining for the case of flow with large particles and quasiequilibrium heterogeneous flow, respectively (see Sect. 1.5). In the case of nonequilibrium heterogeneous flow, where the averaged and fluctuation dynamic and thermal slip occurs between the phases, it is necessary to take into account the contribution by all of the above-identified terms of equations of motion and energy.

We will treat the case where the distributions of averaged velocities and concentrations of the dispersed phase are known. In order to close the set of averaged equations, one must know the turbulent stresses of gas $\overline{u_i'u_j'}$ and the turbulent heat flux $\overline{u_j't'}$, as well as the correlations of the fluctuations of particle concentration with the fluctuations of gas velocity and temperature $\overline{\varphi'u_i'}$ and $\overline{\varphi't'}$ which may be represented as follows [14, 15]:

$$\overline{\varphi'u_i'} = -\tau_p g_p \overline{u_i'u_j'}\frac{\partial \Phi}{\partial x_j}, \tag{2.50}$$

$$\overline{\varphi't'} = -\tau_p g_{pt} \overline{u_j't'}\frac{\partial \Phi}{\partial x_j}, \tag{2.51}$$

where

$$g_p = T_{pL}/\tau_p - 1 + \exp(-T_{pL}/\tau_p), \quad g_{pt} = T_{pLt}/\tau_p - 1 + \exp(-T_{pLt}/\tau_p).$$

One can subtract (2.47)–(2.49) from (2.44)–(2.46), respectively, and derive the fluctuation equations of continuity, motion, and energy of the gas phase in the presence of particles,

$$\sum_j \frac{\partial u_j'}{\partial x_j} = 0, \tag{2.52}$$

$$\frac{\partial u_i'}{\partial \tau} + \sum_j \left[u_j' \frac{\partial U_i}{\partial x_j} + U_j \frac{\partial u_i'}{\partial x_j} + \frac{\partial (u_i' u_j')}{\partial x_j} \right] = -\frac{1}{\rho} \frac{\partial p'}{\partial x_i} + \nu \sum_j \frac{\partial^2 u_i'}{\partial x_j \partial x_j}$$

$$+ \sum_j \frac{\partial (\overline{u_i' u_j'})}{\partial x_j} - \frac{\rho_{\mathrm{p}} \Phi}{\rho} \frac{(u_i' - v_i')}{\tau_{\mathrm{p}}} - \frac{\rho_{\mathrm{p}} \varphi'}{\rho} \frac{[(U_i - V_i) + (u_i' - v_i')]}{\tau_{\mathrm{p}}} + \frac{\rho_{\mathrm{p}} \overline{\varphi' u_i'}}{\rho \tau_{\mathrm{p}}},$$

$$(2.53)$$

$$\frac{\partial t'}{\partial \tau} + \sum_j \left[u_j' \frac{\partial T}{\partial x_j} + U_j \frac{\partial t'}{\partial x_j} + \frac{\partial (u_j' t')}{\partial x_j} \right] = a \sum_j \frac{\partial^2 t'}{\partial x_j \partial x_j} + \sum_j \frac{\partial (\overline{u_j' t'})}{\partial x_j}$$

$$- \frac{C_{\mathrm{p_p}} \rho_{\mathrm{p}} \Phi}{C_{\mathrm{p}} \rho} \frac{(t' - t_{\mathrm{p}}')}{\tau_t} - \frac{C_{\mathrm{p_p}} \rho_{\mathrm{p}} \varphi'}{C_{\mathrm{p}} \rho} \frac{[(T - T_{\mathrm{p}}) + (t' - t_{\mathrm{p}}')]}{\tau_t} + \frac{C_{\mathrm{p_p}} \rho_{\mathrm{p}} \overline{\varphi' t'}}{C_{\mathrm{p}} \rho \tau_t}. \quad (2.54)$$

The fluctuation equation of continuity (2.50) has a form similar to that of the respective (1.9) for a single-phase flow. Equations (2.53) and (2.54) differ from analogous equations of motion and energy for single-phase gas (1.10) and (1.11) by the presence in their right-hand parts of terms which take into account the dynamic and thermal effect of the dispersed phase on the carrier flow. These equations indicate that the inverse effect of particles on the fluctuation motion and heat transfer of carrier gas is defined by the fluctuation and averaged dynamic and thermal slip of the dispersed phase, as well as by the fluctuations of the particle concentration. Note that the contribution made by the penultimate terms of the right-hand parts of (2.53) and (2.54) will be determining for the case of flow with large particles characterized by a significant difference of the averaged velocities and temperatures between the phases.

We will derive the equation for the second moments of fluctuations of velocity of the carrier phase in the presence of particles by analogy with the case of single-phase flow in Sect. 1.2. We will first replace j by k in (2.53) for u_i' and multiply both parts of the resultant equation by u_j',

$$u_j' \frac{\partial u_i'}{\partial \tau} + \sum_k \left[u_j' u_k' \frac{\partial U_i}{\partial x_k} + u_j' U_k \frac{\partial u_i'}{\partial x_k} + u_j' \frac{\partial (u_i' u_k')}{\partial x_k} \right]$$

$$= -u_j' \frac{1}{\rho} \frac{\partial p'}{\partial x_i} + u_j' \nu \sum_k \frac{\partial^2 u_i'}{\partial x_k \partial x_k} + u_j' \sum_k \frac{\partial (\overline{u_i' u_k'})}{\partial x_k} - \frac{\rho_{\mathrm{p}} \Phi}{\rho} \frac{u_j' (u_i' - v_i')}{\tau_{\mathrm{p}}}$$

$$- \frac{\rho_{\mathrm{p}} \varphi'}{\rho} \frac{u_j' [(U_i - V_i) + (u_i' - v_i')]}{\tau_{\mathrm{p}}} + \frac{\rho_{\mathrm{p}} u_j' \overline{\varphi' u_i'}}{\rho \tau_{\mathrm{p}}}. \quad (2.55)$$

We will write a similar equation for u'_j and multiply both its parts by u'_i,

$$u'_i \frac{\partial u'_j}{\partial \tau} + \sum_k \left[u'_i u'_k \frac{\partial U_j}{\partial x_k} + u'_i U_k \frac{\partial u'_j}{\partial x_k} + u'_i \frac{\partial (u'_j u'_k)}{\partial x_k} \right]$$

$$= -u'_i \frac{1}{\rho} \frac{\partial p'}{\partial x_j} + u'_i \nu \sum_k \frac{\partial^2 u'_j}{\partial x_k \partial x_k} + u'_i \sum_k \frac{\partial \overline{(u'_j u'_k)}}{\partial x_k}$$

$$- \frac{\rho_p \Phi}{\rho} \frac{u'_i (u'_j - v'_j)}{\tau_p} - \frac{\rho_p \varphi'}{\rho} \frac{u'_i[(U_j - V_j) + (u'_j - v'_j)]}{\tau_p} + \frac{\rho_p u'_i \varphi' u'_j}{\rho \tau_p}. \quad (2.56)$$

We will combine (2.55) and (2.56) term-by-term and perform averaging. As a result, the equation of transport of turbulent stresses of gas in the presence of particles takes the form:

$$\frac{\partial \overline{(u'_i u'_j)}}{\partial \tau} + \sum_k U_k \frac{\partial \overline{(u'_i u'_j)}}{\partial x_k} = \sum_k \frac{\partial}{\partial x_k} \left[\nu \frac{\partial \overline{(u'_i u'_j)}}{\partial x_k} - \overline{u'_i u'_j u'_k} \right]$$

$$- \sum_k \left[\overline{(u'_j u'_k)} \frac{\partial U_i}{\partial x_k} + \overline{(u'_i u'_k)} \frac{\partial U_j}{\partial x_k} \right] - \frac{1}{\rho} \left(\overline{u'_i \frac{\partial p'}{\partial x_j}} + \overline{u'_j \frac{\partial p'}{\partial x_i}} \right) - 2\nu \sum_k \overline{\frac{\partial u'_i}{\partial x_k} \frac{\partial u'_j}{\partial x_k}}$$

$$- \frac{\rho_p}{\rho \tau_p} \left[\Phi(2\overline{u'_i u'_j} - \overline{v'_i u'_j} - \overline{u'_i v'_j}) + \overline{\varphi' u'_j}(U_i - V_i) \right.$$

$$\left. + \overline{\varphi' u'_i}(U_j - V_j) + (2\overline{\varphi' u'_i u'_j} - \overline{\varphi' v'_i u'_j} - \overline{\varphi' u'_i v'_j}) \right]. \quad (2.57)$$

Equation (2.57) differs from the similar equation for single-phase gas (1.14) by the presence in the right-hand part of the last group of terms which take into account the dynamic effect of the dispersed phase on the carrier flow. The inverse effect of particles on the balance of Reynolds stresses of carrier gas is caused by the fluctuation and averaged slip of the dispersed phase, as well as by the fluctuations of particle concentration.

The set of (2.47), (2.48), (2.50), and (2.57) turns out to be nonclosed, because (2.57) includes unknown triple correlations of fluctuations of the carrier phase velocities, as well as the correlations associated with the fluctuations of concentration and velocity of the dispersed phase. Various models are used to derive the closed set of equations describing the averaged motion of gas in the presence of particles. Most extensively employed (similar to the theory of turbulent single-phase flows) are algebraic, one-parameter, and two-parameter models.

2.4.1 Algebraic Models

The concepts of the Prandtl semiempirical theory of turbulence are usually used in models of this type (see Sect. 1.2). In his pioneering study, Abramovich

[1] used the mixing length theory to determine the fluctuation velocities of gas and particles. The thus developed model is based on the equation of conservation of momentum of turbulent eddy and particles moving in this eddy, as well as on the equation of fluctuation motion of particles within the eddy. It is assumed that that low-inertia particles are entrained in the fluctuation motion by turbulent eddies of the carrier phase; as a result, the fluctuation velocity of gas decreases. The obtained values of fluctuation velocities of gas and particles are used to find correlations by multiplying together the respective fluctuation quantities, which makes this method very approximate. Models of this type were developed further in [2–5, 24, 27, 29, 50, 59].

2.4.2 One-Parameter Models

The widest acceptance (similar to the case of single-phase flow) was received by the model based on the equation for turbulent energy.

In order to construct the equation of transport of turbulent energy of gas in the presence of particles, the equation of fluctuation motion (2.53) must be multiplied by u_i', summed over i, and then averaged. The resultant equation will have the form

$$
\frac{\partial k}{\partial \tau} + \sum_j U_j \frac{\partial k}{\partial x_j} = \sum_j \frac{\partial}{\partial x_j} \left[\nu \frac{\partial k}{\partial x_j} - \overline{u_j' \left(\frac{1}{2} \sum_i u_i'^2 + \frac{p'}{\rho} \right)} \right]
$$

$$
- \sum_j \sum_i \overline{u_i' u_j'} \frac{\partial U_i}{\partial x_j} - \nu \sum_j \sum_i \overline{\frac{\partial u_i'}{\partial x_j} \frac{\partial u_i'}{\partial x_j}}
$$

$$
- \sum_i \frac{\rho_{\mathrm{p}}}{\rho \tau_{\mathrm{p}}} \left[\Phi(\overline{u_i' u_i'} - \overline{u_i' v_i'}) + \overline{\varphi' u_i'}(U_i - V_i) \right.
$$

$$
\left. + (\overline{\varphi' u_i' u_i'} - \overline{\varphi' u_i' v_i'}) \right]. \tag{2.58}
$$

In accordance with the equation of transport of turbulent energy of single-phase gas (1.24), (2.58) may also be rewritten in a condensed form,

$$
\frac{Dk}{D\tau} = D + P - \varepsilon - \varepsilon_{\mathrm{p}}, \tag{2.59}
$$

where the additional dissipation ε_{p} caused by the presence of particles has the form:

$$
\varepsilon_{\mathrm{p}} = \sum_i \frac{\rho_{\mathrm{p}}}{\rho \tau_{\mathrm{p}}} \left[\Phi(\overline{u_i' u_i'} - \overline{u_i' v_i'}) + \overline{\varphi' u_i'}(U_i - V_i) + (\overline{\varphi' u_i' u_i'} - \overline{\varphi' u_i' v_i'}) \right]. \tag{2.60}
$$

The terms on the right-hand side of (2.60) are responsible for the dissipation of turbulent energy caused by the fluctuation interphase slip, the correlation of the fluctuations of particle concentration with the fluctuation

velocity of carrier gas, and the presence of averaged dynamic slip, as well
as by the correlations of the fluctuations of particle concentration and the
fluctuation velocities of the phases, respectively.

The authors of a number of studies (for example, [17, 20, 21]) tried to
estimate the terms in the right-hand part of (2.60) for particle-laden flows
of different types. It was demonstrated that, in flows with relatively inertial
particles ($Stk_L \geq 1$), the fluctuations of concentration of the dispersed phase
do not correlate with the field of fluctuation velocity of gas. This implies
the smallness of the second and third terms of the right-hand part of (2.60)
compared to its first term. Therefore, in the case of the quasiequilibrium
and nonequilibrium flows (see Table 1.1), the first term on the right-hand
side of (2.60) will play the determining part in the process of dissipation of
turbulence. In the case of a flow with large particles which are not entrained
in the fluctuation motion by energy-carrying eddies of the carrier phase, the
expression for ε_p may be written as:

$$\varepsilon_p = \sum_i \frac{\rho_p \Phi}{\rho \tau_p} \overline{u_i' u_i'} = \frac{2Mk}{\tau_p}. \tag{2.61}$$

Note that, in the case of a flow with large particles whose relaxation time
is significant, the value of additional dissipation of turbulent energy will be
negligible compared to other terms of (2.58).

As was demonstrated by the experimental results, the presence of large
particles in the flow may cause additional generation (production) of turbulence of the carrier gas. This mechanism is in no way taken into account in
writing (2.45). We will write (2.59) as:

$$\frac{Dk}{D\tau} = D + P - \varepsilon + P_p - \varepsilon_P, \tag{2.62}$$

where P_p is the term responsible for the additional production of turbulent
energy because of the presence of the dispersed phase. Therefore, the inclusion
of the modification of turbulence in heterogeneous flows presumes a correct
description of the terms of (2.62) responsible for the generation (P_p) and
dissipation (ε_p).

A mathematical model is given in Sect. 4.3, which describes the processes
of additional dissipation of turbulence by low-inertia particles (quasiequilibrium flow) and of additional generation of turbulence in wakes behind moving
particles (flow with large particles). Analysis was performed in a diffusionless
(algebraic) approximation, i.e., disregarding the contribution by the diffusion
term D to (2.62). The effect of particles on the steady-state hydrodynamically developed pipe flow is treated; for this flow, the left-hand part of (2.62)
goes to zero. In addition, for the purpose of deriving simple analytical relations, analysis was made for moderate values of particle concentration, when
the effect of particles on the distribution of averaged velocity of the carrier
gas was minor. As a result, expressions for the source terms ε_p and P_p were

derived and used to find two complexes of physical parameters responsible for the dissipation and generation of turbulent energy of the carrier gas under conditions of quasiequilibrium flow and flow with large particles, respectively.

Good agreement between the calculation results and the available experimental results leads one to expect the efficiency of the model in the case of a nonequilibrium flow, when the joint action is possible of both mechanisms (laminarizing and turbulizing) of the effect of particles on turbulence.

2.4.3 Two-Parameter Models

As in studying single-phase turbulent flows, the most generally employed model is the two-parameter k–ε model of turbulence with the equation for the rate of dissipation used as the second equation.

By analogy with (1.28) for a single-phase flow, we have, in the case of a particle-laden flow,

$$\frac{D\varepsilon}{D\tau} = D_\varepsilon + P_\varepsilon - \varepsilon_\varepsilon - \varepsilon_{\varepsilon p}, \qquad (2.63)$$

where $\varepsilon_{\varepsilon p}$ is the decrease in dissipation because of the presence of particles.

The expression for $\varepsilon_{\varepsilon p}$ is most commonly represented in the form [18, 36]

$$\varepsilon_{\varepsilon p} = C_{\varepsilon 3} \frac{\varepsilon}{k} \varepsilon_p, \qquad (2.64)$$

where the constant $C_{\varepsilon 3}$ may take the following values: $C_{\varepsilon 3} = 1.0$ [33], $C_{\varepsilon 3} = 1.2$ [18], and $C_{\varepsilon 3} = 1.9$ [7].

2.4.4 Methods of Direct Numerical Simulation

In conclusion, we must dwell briefly on the methods of direct numerical simulation (DNS) which are rapidly developing in recent years. A method of direct numerical simulation is the solution of nonstationary Navier–Stokes equations for instantaneous values without involving additional closing relations or equations, i.e., actually without the simulation of turbulence. The well-known limitation of such a method is the impossibility of using it at moderate or high values of the Reynolds number. A variety of this method is the method of large eddy simulation (LES) which involves the treatment of only large energy-carrying eddies [30]. In this manner, an attempt is made at obviating the disadvantage identified earlier and extending the range of application of the method.

In the overwhelming majority of early investigations of particle-laden two-phase flows [16, 44, 53], these two methods were used to simulate the motion of single particles; in accordance with the classification of heterogeneous flows developed in Sect. 1.5, this corresponds to the case of a weakly dusty flow without the inverse effect of particles on the carrier gas parameters. These investigations were performed to study the behavior of particles. For this purpose, the trajectories of a large ensemble of particles introduced into a turbulent flow were calculated, which was followed by the averaging of the obtained spatial characteristics of particle motion. Note that the spatial resolution was much less than the particle size proper. In performing the calculations, it was not intended to determine the parameters of gas flow about a particle. This was not necessary, because the particle motion is calculated in the usual way, i.e., using the law of resistance of the dispersed phase. The particle drag is defined by the Reynolds number; for determining the value of this number, one needs to know the carrier gas velocity rather than the distribution of this velocity over the particle contour. The foregoing restriction in the calculation of particle motion is valid only when describing the behavior of very fine particles whose size is less than the size of the smallest turbulent eddies (Kolmogorov scale).

In more recent investigations [6, 8, 9, 45, 47], the methods of direct numerical simulation were used as advantage for the calculation of weakly dusty flows with the inverse effect of particles on the characteristics of flow of the carrier phase. In this case, the calculations are performed in several iteration steps. First, the parameters of motion of "pure" gas are calculated. For this purpose, it is usually assumed that the fluctuations of gas velocity obey the normal law. In the known field of gas velocities, the trajectories of particles are calculated by integrating the equations of their motion. Then, given a fairly representative ensemble of particles, one finds the averaged characteristics of the dispersed phase which are then used to calculate the gas phase flow in the next stage. The thus obtained "new" field of gas velocities serves a basis for the calculations of particle trajectories at the next iteration step, and so on. The calculations are performed until the difference between the obtained characteristics of motion of both phases of heterogeneous flow at the previous and subsequent iteration steps is within the preassigned error.

3

Physical Simulation
of Particle-Laden Gas Flows

3.1 Preliminary Remarks

The physical simulation and mathematical simulation of heterogeneous flows pursue one and the same objective, that of constructing the theory of multi-phase flows. This objective may apparently be achieved by using separately either the experimental or the computational methods of investigation. At the same time, it is apparent that each one of the simulation methods is charac-terized by a number of inherent advantages and disadvantages. Therefore, the process of construction of the theory will be more effective if the means of physical and mathematical simulation are mutually complementary. Possible ways of such interaction are obvious. For example, measurement results are extensively used to verify the mathematical models being developed. Note the importance of experimental data essential for the formulation of the boundary conditions for the dispersed phase of a heterogeneous flow. In turn, the calcula-tion results are called upon to minimize the required volume of experimental data and to contribute to a more profound interpretation of measurement results.

In this chapter, the fundamentals of physical simulation of turbulent flows with the dispersed phase in the form of solid particles will be treated using an intensively developing optical diagnostic method, i.e., laser Doppler anemometry. The probe and photographic means of investigation of disperse flows, which have been used for decades, are described in detail in mono-graphs [5, 10, 12].

Laser Doppler anemometers (LDA) have been used for more than two decades to investigate single-phase flows. For measuring the kinematic char-acteristics of flow of a continuous medium, this medium is injected with tracer particles of submicron or micron sizes whose mass and volume concentration is negligible. If certain conditions are met, the instantaneous velocities of the tracer particles will be almost equal to the respective velocities of the con-tinuous carrier medium. Heterogeneous flows, in which particles (droplets, bubbles) are naturally present, are likewise investigated using an LDA. It is safe to say that the LDA has by now become the most powerful (and often the only) means of local diagnostics of such flows.

Section 3.2 deals briefly with the fundamentals and advantages of the method of laser Doppler anemometry. In spite of all numerous advantages of the method of laser Doppler anemometry, its use for studying turbulent heterogeneous flows is predetermined by the solution of a number of specific procedural problems. Special aspects and objectives of experimental investigation of particle-laden gas flows are described in Sect. 3.3. Various aspects of study of the behavior of particles and of their inverse effect on the parameters of carrier gas flow are treated in Sects. 3.4 and 3.5, respectively. In Sect. 3.6, examples are given of experimental facilities for the investigation of turbulent heterogeneous flows.

3.2 Laser Doppler Anemometry and its Advantages

The method of laser Doppler anemometry is one of the optical diagnostic methods which are extensively used to investigate flows [4, 6, 7, 9, 11, 16, 17, 20–23, 25, 26, 31]. The key advantage of the optical diagnostic methods is the possibility of performing measurements without disturbance of the flow in the region being investigated. Along with this common advantage of the optical diagnostic methods, the method of laser Doppler anemometry exhibits the following unique features which made it in recent years a powerful tool for studying the fine structure of flows:

1. A high spatial resolution owing to the smallness of the measuring volume
2. A high time resolution because of the combination of the small measuring volume with a high-speed processor of the Doppler signal (this enables one to perform measurements of instantaneous values of velocity)
3. Freedom from the need to perform calibration owing to the absolutely linear correlation between the Doppler signal frequency and velocity
4. The possibility of performing multicomponent measurements, i.e., measurements of one, two, and three components of the velocity vector
5. The pointed directivity of measurements, because the quantity being measured is the projection of the velocity vector in the direction defined by the optical system
6. The possibility of investigating flows with reversible velocity, as well as of performing measurements at a velocity close to zero under conditions of the optoelectronic shift of frequency
7. A high stability and repetition of measurement results owing to the stability and linearity of optical electromagnetic waves.

The necessary conditions in using the LDA include the following requirements:

Fig. 3.1. A scheme of shaping the interference (measuring) volume of laser Doppler anemometer (LDA)

1. A low optical density of the medium being investigated
2. The presence in the flow of particles which are light-scattering centers
3. An access for laser beams via windows or an inlet for an optical probe.

All of the advantages of the method of laser Doppler anemometry identified earlier made it in recent years a powerful tool for the diagnostics of single-phase and heterogeneous flows.

We will consider later the fundamentals of shaping the interference volume of LDA, as well as a signal from a light-scattering particle.

Two laser beams with Gaussian distribution of intensity (TEM$_{00}$-mode) and wavelength λ, which intersect at angle θ, shape the interference volume in the intersection region (Fig. 3.1). The interference volume is an ellipsoid of revolution with axes (the intensity level e^{-2}),

$$d_x = \frac{d_f}{\cos(\theta/2)}, \ d_y = d_f, \ d_z = \frac{d_f}{\sin(\theta/2)}, \tag{3.1}$$

where the diameter of focused laser beam at the waist is defined as:

$$d_f = \frac{4f\lambda}{\pi E d_1}. \tag{3.2}$$

Here, f is the focal distance of the front lens, E is the beam expansion coefficient, and d_1 is the laser beam diameter before expansion.

Therefore, the minor axis of the ellipsoid (dimensions d_x and d_y) depends on the extent of the laser beam waist and varies weakly with the beam intersection angle, while the major axis (dimension d_z) increases with decreasing angle between the beams.

The interference lattice spacing is defined by the relation

$$\delta_f = \frac{\lambda}{2\sin(\theta/2)}. \tag{3.3}$$

A scheme of the interference volume is given in Fig. 3.2.

In the general case, the measuring volume is a region whose size is smaller or larger than the size of the interference volume and is defined, first of all, by the parameters of sensitivity of the transmitting and receiving

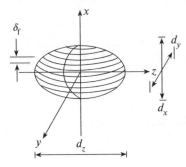

Fig. 3.2. A scheme of the interference (measuring) volume of LDA

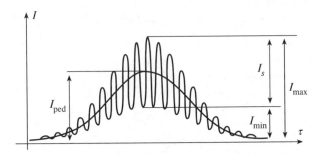

Fig. 3.3. A Doppler signal from a light-scattering particle

optoelectronic systems (laser radiation power, voltage on the photomultiplier, electronic amplification of signal) and by the characteristics of light-scattering particles (size, optical properties, concentration).

A particle scatters light while crossing the interference fringes (Fig. 3.1). The intensity of scattered radiation will vary in accordance with the variation of light intensity in the region of laser beam intersection. The signal from the particle that comes from the photomultiplier consists of two parts (Fig. 3.3), namely, (1) a low-frequency part or Gaussian pedestal with the peak amplitude I_{ped} produced by light scattered by the particle from both laser beams, and (2) a high-frequency part or sinusoidal signal with Gaussian distribution and of amplitude I_s produced by the interference of light scattered by the particle from both beams.

One of the basic characteristics of a Doppler signal is its visibility determined from the following relation:

$$\eta = \frac{I_{\max} - I_{\min}}{I_{\max} + I_{\min}}. \tag{3.4}$$

The signal visibility depends on number of factors which include:

1. The size of the light-scattering particle and its optical properties
2. The polarization properties and intensity ratio of the beams which shape the interference volume

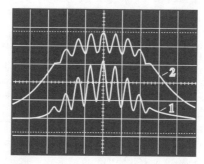

Fig. 3.4. A typical form of Doppler signals: (1) signal from a small particle, (2) signal from a large particle

3. The angle of beam intersection
4. The characteristics and location of receiving optics

One can readily demonstrate that an increase in the particle size (all other things being equal) causes a decrease in the signal visibility. Indeed, one can see from simple physical considerations that the low-frequency component of the signal increases with increasing particle size, while the high-frequency component, on the contrary, decreases. For example, in the case of small particles, $I_{max}/I_{min} \to \infty$, and, in accordance with (3.4), we have $\eta \to 1$. For large particles, $I_{max}/I_{min} \to 1$ and $\eta \to 0$. A typical form of Doppler signals is given in Fig. 3.4.

3.3 Special Features and Objectives of Experimental Studies of Heterogeneous Flows

Measurements always entail errors on whose magnitude the measurement accuracy depends. In the case of LDA investigations of single-phase flows, it is possible to monitor the accuracy of measurements by their duplication using, for example, a hot-wire anemometer. In addition, there is always a possibility of comparing the results with the available literature data by other authors for single-phase flows of "canonical" forms. As to LDA investigations of heterogeneous flows, no such possibilities exist for the following two reasons:

1. At present, the LDA is in fact the only means of local diagnostics of heterogeneous flows, and this rules out the possibility of duplication of measurements using other experimental means
2. The extreme complexity of heterogeneous flows, which consists in a significantly larger number of dimensionless determining parameters (compared to single-phase flow), resulted in that the data of different researchers actually defy comparison and systematization.

In support of the foregoing, we will cite only one illustrative example which characterizes the current status of investigations of dust-laden flows.

Lee and Durst [11] used the LDA to study the effect of glass particles of different sizes on the characteristics of carrier air for an upward pipe flow at $U_{xc} = 5.7\,\mathrm{m\,s^{-1}}$. The measurement results showed the velocity of large particles of glass ($d_p = 800\,\mu\mathrm{m}$) is almost constant over the pipe cross-section and is $V_x = 1.4\,\mathrm{m\ s^{-1}}$. However, simple estimates indicate that the free-fall velocity of such particles exceeds the air velocity. As a result, this upward flow must not entrain such large particles. Therefore, the value of the velocity of large particles obtained in [11] must be viewed with criticism.

In view of the foregoing, it is clear that correct LDA measurements in heterogeneous flows assume special urgency.

The study of heterogeneous flows pursues the objective of solving problems of two main classes, namely:

1. The investigation of the behavior of particles of the dispersed phase moving in a flow of gas (liquid). This presumes performing measurements of the dimensions of disperse inclusions (in the case of polydisperse flow), as well as of the fields of instantaneous velocities and concentrations of particles.
2. The investigation of the inverse effect of the dispersed phase on the flow characteristics of the carrier phase. For this purpose, one must perform measurements of the fields of instantaneous velocities of tracer particles (which simulate the motion of continuous medium) in the presence of the dispersed phase followed with statistical treatment of the results.

The instantaneous values of velocities of particles and gas phase, obtained as a result of measurements, are used to find the required statistical characteristics. More often than not this reduces to determining the averaged values of velocities and their mean-square deviations, according to the following relations:

$$V_j = \frac{1}{N}\sum_i^N v_{ji}, \tag{3.5}$$

$$\sqrt{\overline{v_j'^2}} = \sqrt{\frac{1}{N}\sum_i^N (v_{ji} - V_j)^2}, \tag{3.6}$$

where V_j and $\sqrt{\overline{v_j'^2}}$ are the jth components of the averaged velocity of particles and its mean-square deviation, respectively; v_{ji} is the ith value of measured particle velocity; and N is the number of measurements in the ensemble over which the averaging is performed.

The respective characteristics of motion for gas are found similarly,

$$U_j = \frac{1}{N}\sum_i^N u_{ji}, \tag{3.7}$$

$$\sqrt{\overline{u_j'^2}} = \sqrt{\frac{1}{N} \sum_{i}^{N} (u_{ji} - U_j)^2},\tag{3.8}$$

where U_j and $\sqrt{\overline{u_j'^2}}$ are the jth components of the averaged velocity of gas and its mean-square deviation, u_{ji} is the ith value of measured gas velocity, and N is the number of measurements in the ensemble.

Some aspects of the solution of problems of the basic classes identified earlier, which are involved in the investigation of heterogeneous flows, will be treated later.

3.4 Special Features of Studies of the Behavior of Solid Particles

The main objective of LDA investigations of the characteristics of motion of large particles representing the dispersed phase of heterogeneous flow is to measure the instantaneous velocities of the particles. Strictly speaking, commercially available LDAs (see Fig. 3.5) are intended for measurement of the velocities of small tracer particles which simulate the motion of a continuous carrier medium. Because of this, the characteristic features of light scattering must be taken into account in planning measurements of velocities of large particles.

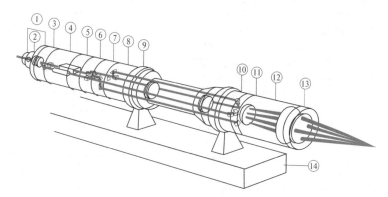

Fig. 3.5. A schematic view of a four-beam two-component LDA manufactured by Dantec (LDA 11): (1) wavelength divider, (2) beam waist adjustment module, (3) neutral beam splitting module, (4) Bragg cell, (5) beam color separation module, (6) beam displacement module, (7) beam color separation module, (8) back-scattering module, (9) support, (10) convergence angle adjustment module, (11) lens holder, (12) beam expander, (13) front lens, and (14) optical bench

3.4.1 Optimization of LDA Parameters

Because the LDA measurements of velocities are based on recording a signal arising when a light-scattering particle crosses a time-varying interference field, the characteristics of radiation scattered by particles must be taken into account in designing the optoelectronic system of LDA and optimizing its parameters. We will consider in brief the characteristic features of light scattering by relatively large particles such as particles of the dispersed phase. This may be done using the Mie scattering theory [34]. The Mie theory shows that the magnitude and angular distribution of the intensity of light depends on the particle diameter. The scattering of light by particles in a heterogeneous flow with sizes of tens and hundreds of micrometers may be analyzed in the so-called large-particle approximation, because the relative size of these particles is much larger than unity, i.e., $\gamma = \pi d_{\mathrm{p}}/\lambda \gg 1$ (d_{p} is the particle diameter and λ is the wavelength of incident light).

The angular distribution of scattered light is often described by dimensionless indicatrix functions which are obtained as a result of normalization of differential scattering cross-sections. The scattering indicatrix of a large transparent particle exhibits the following distinguishing features: all of scattered radiation is largely directed forward, a large number of interference maxima are present, and the smoothed intensity of scattered light does not depend on the particle size.

On constructing the optical scheme of measurements, one must take into account the fact that the scattered radiation of large particles has a diffraction component. In so doing, the diffraction wave does not depend on the particle refractive index and on the incident wave polarization. The diffracted radiation of large spherical particles ($\gamma \gg 1$) is concentrated in a cone of angle $\theta = 3.83/\gamma$ which does not exceed one degree for the majority of particles.

The most important characteristics of the optical system of an LDA are the geometric parameters of the measuring volume. The data of Rinkevichius and Yanina [18] reveal quite a definite correlation between the visibility of the interference pattern and the relative particle size (the ratio of the particle diameter to the interference lattice spacing $d_{\mathrm{p}}/\delta_{\mathrm{f}}$). For example, in measuring the velocities of large particles, the recommended spacing of interference lattice must exceed the particle diameter or be comparable to it. My investigations have revealed that this is not quite true.

An LDA 10 two-channel three-beam laser Doppler anemometer (manufactured by Dantec, Denmark) operating according to a differential scheme was used for measurements. We used only one channel of this LDA for measurements of the axial velocity of particles. The source of radiation in the system is provided by a Russian-made LG-106-ml argon laser. In view of the characteristic features of indicatrices of large particles, all measurements were performed under conditions of forward scattering. An expander (55X12 Beam Expander) and a front lens with a focal distance of 310 mm were used in the optical system: as a result, the measuring volume could be reduced (intensity level e^{-2}) to $0.091 \times 0.091 \times 1.04$ mm, which corresponds to the interference lattice spacing $\delta_f = 2.97\,\mu$m.

The light scattered by particles which find their way into the measuring volume is collected by the receiving optics of photomultipliers (PM), the signal from which is delivered to the Doppler signal processor of the counter type (55L90a LDA Counter Processor). The 55L90a processor manufactured by Dantec operates in the frequency range from 2 kHz to 100 MHz and has a time resolution of 2 ns.

When a 55L90a processor is used in experiments, two characteristics serve as indicators of the quality of the Doppler signal, namely:

1. The data rate (intensity) D_{pV}, kHz
2. The reliability of the incoming signals V, %.

The data rate (intensity) is an extensive quantity, because it depends directly on the concentration of light-scattering particles in the measuring volume and on the sensitivity of the receiving system of the LDA.

The signal reliability (in operation in the "5/8" mode) is defined by the number of particles (in percentage terms), the difference between whose velocities (determined by their time of flight of five and eight spacings of interference lattice) does not exceed the preassigned error (this error was usually 3%). The signal reliability is the main characteristic of its quality and is directly related to the error of measurement. The signal reliability may be reduced for a number of reasons.

Given later are some of these reasons:

1. An inadequate sensitivity of the receiving optoelectronic system of the LDA, which results in low values of the data rate ($D_{pV} = 0\text{--}0.1$ kHz). This is due to the following factors:
   - An inadequate voltage applied to the PM
   - A low number density of light-scattering particles
   - An inadequate amplification of the signal delivered from the PM
   - Weak focusing of the PM optics
2. An excess sensitivity of the optoelectronic system of the LDA, which results in too high values of the data rate ($D_{pV} = 10\text{--}100$ kHz). More often than not, this is caused by the same (but opposing) factors:
   - Too high a voltage applied to the PM
   - A high number density of light-scattering particles
   - An excess amplification of the signal delivered from the PM
3. The operation in too narrow a frequency range preassigned by the high-pass and low-pass filters, as well as by the value of the electronic frequency shift. In this case, it is likely that the probability density curve of velocities (respective frequencies) of particles does not fit the selected frequency range.

Figure 3.6 gives an example of the distributions of particle velocities as functions of signal reliability. In this example, the decrease in signal reliability

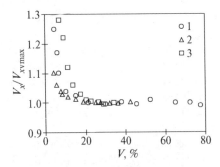

Fig. 3.6. The velocity of large particles being measured as a function of signal quality ($V_{xv\,max}$ is the value of particle velocity obtained for the maximal quality of signal): (1) SiO₂ particles (200 μm), (2) Fe particles (150 μm), and (3) Pb particles (59 μm)

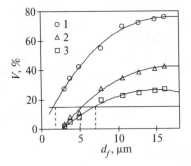

Fig. 3.7. The signal quality as a function of the interference lattice spacing: (1) SiO₂ particles (200 μm), (2) Fe particles (150 μm), and (3) Pb particles (59 μm)

was caused by the excess sensitivity of the optoelectronic system of the LDA because of too high a voltage applied to the PM. One can readily infer from Fig. 3.6 the existence of a minimal value of signal reliability, at which the velocities of large particles may be measured with minimal error. This reliability is $V = 15\%$.

Figure 3.7 gives the distributions of reliability of signals as functions of the interference lattice spacing for some of the particles used. Analysis of these distributions shows that:

1. The signal reliability depends strongly on the interference lattice spacing and on the physical and optical properties of light-scattering particles
2. For each variety of particles, there exists a minimal value of the interference lattice spacing, which corresponds to the lower limit of signal reliability ($V = 15\%$) at which the error of velocity measurement is minimal. For example, for glass particles 200 μm, the minimal spacing of interference lattice is $\delta_{f\,min} = 2\,\mu m$, and for particles of iron 150 μm in diameter and lead 59 μm in diameter, $\delta_{f\,min} = 5\,\mu m$ and $7\,\mu m$, respectively

3. An increase in the interference lattice spacing results in a higher signal quality; a value of the interference lattice spacing exists above which almost no change of signal reliability is observed. For the particles used in the experiments, this value of the interference lattice spacing is in the range $\delta_{\mathrm{f}} = 12\text{--}16\,\mu\mathrm{m}$.

One must bear in mind that an increase in the lattice spacing causes an increase in the measuring volume, thereby reducing the locality of measurements. This fact assumes special importance just in measuring the velocities of large particles for which the size of the "effective" measuring volume may exceed significantly that of the measuring (interference) volume determined by the intensity level e^{-2}. The method of approximate estimation of the value of "effective" measuring volume for large light-scattering particles was described in [33].

3.4.2 Measurement of the Velocities of Polydisperse Particles

Because the overwhelming majority of flows contain significantly polydisperse particles, it is this particular fact that has a determining effect on the shaping of the statistical properties of the dispersed phase. In this subsection, we treat the basics of the procedure of correct measurements of instantaneous velocities of polydisperse particles using LDA. This problem assumes special urgency because of the specific features (identified in Sect. 3.3) arising during LDA investigations of heterogeneous flows.

Figure 3.8 gives a typical mass distribution of particles in the form of a histogram obtained using sieve analysis. Note that a strong correlation exists between the mass distribution of particles and their number distribution [27]. For the example at hand, one can readily demonstrate that the respective distributions of particles actually coincide. Therefore, in further analysis, we will restrict ourselves to treating the mass distribution alone. In order to find the probability density function (PDF) of particle sizes, we will determine the

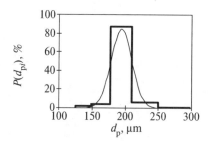

Fig. 3.8. A histogram of the mass distribution of spherical SiO_2 particles with a rated diameter of $200\,\mu\mathrm{m}$ (the curve indicates the approximating PDF of particle sizes)

average diameter of particles and its mean-square deviation. We assume an equiprobable distribution of particles within each one of N fractions to derive

$$Md_\mathrm{p} = \sum_{1}^{N} d_{\mathrm{p}_i} P(d_{\mathrm{p}_i}), \tag{3.9}$$

where Md_p is the mathematical expectation of the particle diameter, d_{p_i} is the average value of diameter of particles of the ith fraction, $P(d_{\mathrm{p}_i})$ is the probability that a particle belongs to the ith fraction, and $\sum_{1}^{N} P(d_{\mathrm{p}_i}) = 1$ is the normalization condition.

Given the average diameter of particles, the dispersion may be determined from the relation

$$Dd_\mathrm{p} = \sum_{1}^{N} (d_{\mathrm{p}_i} - Md_\mathrm{p})^2 P(d_{\mathrm{p}_i}). \tag{3.10}$$

The mean-square deviation of particle diameters is related to dispersion as:

$$\sigma = \sqrt{Dd_\mathrm{p}}. \tag{3.11}$$

Table 3.1 gives the data necessary for the calculation of mathematical expectation and mean-square deviation of particle diameters. By using relations (3.9)–(3.11), one can readily obtain $Md_\mathrm{p} = 194.1\,\mu\mathrm{m}$ and $\sigma = 15.64\,\mu\mathrm{m}$.

For further analysis, we will assume that the particle distribution obeys the normal law. In this case, the statistical parameters of particle sizes found earlier, namely, Md_p and σ fully define their PDF.

In performing probability calculations, it is convenient to use the dimensionless PDF which has the following form for the normal law of distribution:

$$f(\bar{d}_\mathrm{p}) = f_\mathrm{max} \exp\left[-\frac{(\bar{d}_\mathrm{p} - M\bar{d}_\mathrm{p})^2}{2\bar{\sigma}^2} \right], \tag{3.12}$$

where $f_\mathrm{max} = \frac{1}{\sqrt{2\pi}\bar{\sigma}}$, $\bar{d}_\mathrm{p} = \frac{d_\mathrm{p}}{A}$, $\bar{\sigma} = \frac{\sigma}{A}$, and $\int_{0}^{\infty} f(\bar{d}_\mathrm{p}) d(\bar{d}_\mathrm{p}) = 1$ is the normalization condition.

Table 3.1. Parameters of the distribution of glass particles with a rated diameter of 200 μm

| fraction no. | Characteristics of particle distribution | | |
	$d_{\mathrm{p}_i}\,(\mu\mathrm{m})$	$\Delta d_{\mathrm{p}_i}\,(\mu\mathrm{m})$	$P(d_{\mathrm{p}_i})$
1	137	24	0.02
2	163	28	0.04
3	193.5	33	0.87
4	230	40	0.06
5	273.5	47	0.01

In order to reduce the concrete distribution $f(d_p)$ to the form of (3.12), it is necessary to determine the normalizing constant A. One can easily do this using the following relation:

$$A = \sum_1^N P(d_{p_i})\Delta d_{p_i},\qquad(3.13)$$

where Δd_{p_i} is the difference between the maximal and minimal diameters of particles of the ith fraction. We use the data of Table 3.1 to obtain $A = 33.18\,\mu\text{m}$ from (3.13). Given the constant, we determine the average diameter of particles and its mean-square deviation in a dimensionless form in accordance with (3.12) as $M\bar{d}_p = 5.850$ and $\bar{\sigma} = 0.4714$. The expression for the dimensionless PDF of particle sizes takes the form:

$$f(\bar{d}_p) = \frac{1}{\sqrt{2\pi}\,0.4714}\,\exp\left[-\frac{(\bar{d}_p - 5.850)^2}{2 \times 0.4714^2}\right].\qquad(3.14)$$

The PDF of particle sizes is given in Fig. 3.8.

Determination of Mathematical Expectation of the Particle Diameter and Its Mean-Square Deviation

In order to estimate the averaged velocities of particles and their mean-square deviations using the procedure described in [28], it is necessary to know the average diameters of particles and their deviations. In performing measurements of particle velocities at some fixed values of LDA sensitivity (defined primarily by the value of voltage applied to the photomultiplier – PM), we investigate in fact the velocity of particles whose distribution is truncated on the left rather than the velocity of the entire set of polydisperse particles whose distribution obeys the normal law. As a result, the measured values of statistical characteristics of particle motion turn out to be different from their actual values (this fact will be analyzed later).

The truncated normal distribution in the general case is defined by four parameters [1], namely, (1) the mathematical expectation of initial normal distribution $M\bar{d}_p$, (2) the mean-square deviation of the initial normal distribution $\bar{\sigma}$, (3) the point of truncation on the left $\bar{d}_{p1}$, and (4) the point of truncation on the right $\bar{d}_{p2}$.

The expressions for mathematical expectation $M_0\bar{d}_p$ and mean-square deviation $\bar{\sigma}_0$ of truncated normal distribution have the form [1]:

$$M_0\bar{d}_p = M\bar{d}_p + B\bar{\sigma},\qquad(3.15)$$

$$\bar{\sigma}_0 = \bar{\sigma}\sqrt{1 - B^2 - C[t_2 f(t_2) - t_1 f(t_1)]},\qquad(3.16)$$

where

$$B = \frac{f(t_1) - f(t_2)}{F(t_2) - F(t_1)}, \quad C = \frac{1}{F(t_2) - F(t_1)}, \quad t_{1,2} = \frac{\bar{d}_{p1,2} - M\bar{d}_p}{\bar{\sigma}},$$

$$f(t_{1,2}) = \frac{1}{\sqrt{2\pi}} \exp(-t^2/2), \quad \text{and} \quad F(t_{1,2}) = \int_0^{t_{1,2}} f(t_{1,2}) dt.$$

We will determine the mathematical expectations and mean-square deviations of the family of truncated normal distributions, which is obtained from initial distribution (3.14). In so doing, we will vary the coordinate of the point of truncation on the left in the range $0 \leq \bar{d}_{p1} < M\bar{d}_p + 6.58\bar{\sigma}$; in view of the value of normalizing constant, this corresponds to the range of particle diameters of $0 \leq d_{p1} < 297\,\mu m$. The maximal coordinate of the point of truncation on the left is taken to be equal to the possible particle size $d_{p\,max} = 297\,\mu m$ (see Fig. 3.8 and Table 3.1). The coordinate of the point of truncation on the left will also be taken to be $\bar{d}_{p2} = M\bar{d}_p + 6.58\bar{\sigma}$.

The distributions of mathematical expectation of particle diameter $M_0 d_p = f_1(d_{p1})$ and its mean-square deviation $\sigma_0 = f_2(d_{p1})$, obtained as a result of calculations, may be further used to determine the averaged particle velocity and its mean-square deviation. The results of calculation of averaged velocities of single glass particles under conditions of their gravity sedimentation in air at rest V_x and their mean-square deviations σ_{0V} are given in Fig. 3.9 in a dimensional form.

In order to evaluate the developed procedure for estimation of averaged velocities and their mean-square deviations, measurements of the respective parameters were performed [30].

Figure 3.10a gives the distribution of averaged velocities of glass particles as a function of the sensitivity of the receiving optoelectronic system of the

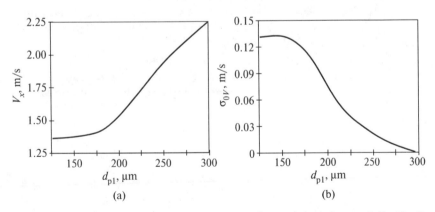

(a)   (b)

Fig. 3.9. The effect of the position of the point of truncation of normal distribution on **(a)** the averaged particle velocity and **(b)** the mean-square deviation of the particle velocity

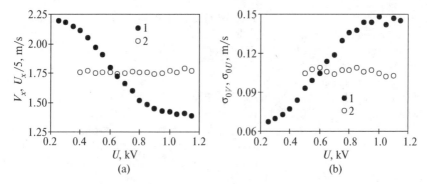

Fig. 3.10. The effect of the sensitivity of the optoelectronic system of LDA on the measured values of (**a**) the averaged velocity and (**b**) the mean-square deviation of the particle velocity: (1) glass (200 μm), (2) glycerin + water (2–5 μm)

LDA, which is defined by the value of voltage applied to the PM. One can see that the obtained values of averaged velocity of particles depend strongly on the LDA sensitivity. This is apparently associated with the fact that the particles used in our investigation are polydisperse. Therefore, in the case of low values of voltage applied to the PM ($U = 0.25$–$0.5\,\text{kV}$), when the amplitude of signals from small particles does not exceed the threshold of sensitivity of the receiving system of the LDA, only the signals from large particles exhibiting high values of velocity under conditions of this investigation are admitted for statistical treatment. When the LDA sensitivity is increased, signals from every smaller particles are admitted. As a result, the averaged particle velocity begins to decrease. In the case of high values of voltage on the PM ($U = 0.8$–$1.15\,\text{kV}$), when signals from the smallest particles are admitted for analysis, the averaged velocity of the latter particles ceases to decrease and reaches its minimal value. The distribution of averaged particle velocity given in Fig. 3.10a agrees well with the similar distribution obtained as a result of analysis (see Fig. 3.9a). Because of this, the minimal value of averaged velocity may be treated as the averaged velocity of the entire set of polydisperse particles. Therefore, the failure to include the effect of sensitivity of the receiving system of LDA on the measured value of averaged velocity of polydisperse particles may lead to significant errors.

Also given in Fig. 3.10a are the results of LDA measurements of the averaged velocity of air under conditions of turbulent downward pipe flow. The measurements were performed at a single point on the pipe axis. In these measurements, light-scattering centers were provided by particles prepared of a mixture of glycerin (50%) and water (50%) using a generator of micron-sized particles (Dantec, Denmark). The size of these tracer particles simulating the carrier air motion was in the range from 2 to 5 μm. Although these particles are significantly polydisperse, one can see in Fig. 3.10a that the averaged velocity of tracer particles does not depend on the sensitivity of the receiving optoelectronic system of LDA. Firstly, this is associated with the fact that the inertia of the tracer particles is so low that the difference between the velocities of the largest and smallest particles does not exceed $10^{-4}\%$ and is within the error of these measurements. Secondly, the absolute difference in size between these particles is insignificant, and the respective range of intensity of light scattered by these particles may lie within one step during variation of the PM voltage ($\Delta U = 0.05\,\mathrm{kV}$).

By their nature, the experimentally observed fluctuations of the velocities of tracer particles are turbulent fluctuations acquired by them in the process of interaction with turbulent eddies of carrier air. As to the mean-square deviation of the velocities of micron-sized particles, shown in Fig. 3.10b, this deviation (as well as the averaged velocity) remains unvaried during the variation of voltage on the PM for the reasons given earlier.

Unlike the case of tracer particles, the measured mean-square deviation of the velocities of large polydisperse glass particles depends largely on the sensitivity of the receiving system of LDA (see Fig. 3.10b). In the case of high values of voltage on the PM ($U = 0.8\text{--}1.15\,\mathrm{kV}$), when signals from the entire set of polydisperse particles are admitted for analysis, the mean-square deviation of the velocities of these particles ceases to increase and reaches its maximal value. The distribution of the fluctuation velocity of glass particles given in Fig. 3.10b agrees well with the similar distribution obtained as a result of theoretical analysis (see Fig. 3.9b). Therefore, the maximal value of the obtained distribution may be treated as an unbiased value of mean-square deviation of the velocities of polydisperse particles. The foregoing leads one to infer that the failure to include the effect of sensitivity of the receiving system of LDA on the measured value of fluctuation of the velocities of polydisperse particles (as well as in the case of measurement of their averaged velocities) may lead to significant errors.

The foregoing is clearly supported by the distributions of the velocities of glass particles, which are given in Fig. 3.11 in the form of histograms obtained for selected values of voltage on the PM.

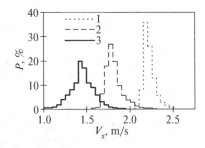

Fig. 3.11. The distribution of the velocities of glass particles for selected values of voltage on the PM: (1) $U = 0.3\,\mathrm{kV}$, (2) $U = 0.6\,\mathrm{kV}$, and (3) $U = 0.9\,\mathrm{kV}$

Heterogeneous flows which occur naturally and are employed in technical devices are often accompanied by physicochemical processes which lead to a variation of the composition of the dispersed phase. Such processes include combustion, phase transitions, coagulation, fragmentation, and so on. This makes obvious the urgency of investigation of flows with particles exhibiting different properties and, consequently, different velocities. The capabilities and limitations of LDA in the investigation of flows carrying bidisperse solid particles were treated in [29]. Three main types of heterogeneous flows with bidisperse particles are known, namely, flows of continuous media carrying solid particles of the same material but of different size, flows with particles of the same size but of different density, and flows of particles of the same size and material but of different "effective" density (hollow particles, porous particles, and the like). The main objective in studying the behavior of bidisperse particles moving in flows is to determine the PDF of their velocities. The obtained PDF of the velocities of particles may be used to find the averaged velocity of the entire plurality of particles, its mean-square deviation, and other necessary statistical parameters. The results of measurements performed in [29] demonstrated the possibility of using LDA for studying the fields of velocities of large bidisperse particles of the same optical properties.

3.4.3 Monitoring of the Accuracy of the Results

One can monitor the validity of the choice of parameters of the measuring system and estimate the error of the results in the investigation of the velocities of large solid particles by way of measuring one of the most important characteristics of particle inertia, namely, the free-fall velocity (settling velocity). The methods of determining this characteristic of particles and the obtained results were described by Gorbis [8]. He determined the settling velocity by measuring the limiting rates of sedimentation of particles in air at rest. According to the principle of relativity of motion, the rate of sedimentation of particles and the velocity of flow suspending them are equal. This is easy to demonstrate.

We will write the equation for the determination of the averaged velocity of motion of a single spherical particle under the effect of the forces of gravity and aerodynamic drag of the medium in the following form:

$$\frac{dV_x}{d\tau} = \frac{U_x - V_x}{\tau_p} \pm g, \tag{3.17}$$

where g is the acceleration of gravity.

We will consider the motion of relatively large non-Stokesian ($1 < Re_p < 1,000$) particles. In this case, the time of dynamic relaxation of a particle has the form

$$\tau_p = \frac{\tau_{p0}}{1 + Re_p^{2/3}/6} \tag{3.18}$$

where $\tau_{p0} = \rho_p d_p^2 / 18\mu$ is the time of dynamic relaxation of a Stokesian particle.

We will derive from (3.17) an expression for determining the settling velocity. For this purpose, we will take into account the fact that $V_x = 0$ and $dV_x/d\tau = 0$. Then, we have

$$U_x = \frac{g\tau_{p0}}{1 + \frac{1}{6}\left(\frac{U_x d_p}{\nu}\right)^{2/3}}. \qquad (3.19)$$

We will now turn to (3.17) with a view to determining the limiting rate of gravity sedimentation of particles in a medium at rest. For this case, we will take into account the fact that $U_x = 0$ and $dV_x/d\tau = 0$. Consequently, (3.17) yields

$$V_x = \frac{g\tau_{p0}}{1 + \frac{1}{6}\left(\frac{V_x d_p}{\nu}\right)^{2/3}}. \qquad (3.20)$$

On analyzing the resultant relations (3.19) and (3.20) for the same physical properties of the medium and particles, one can infer that the limiting rate of sedimentation of particles and the settling velocity are equal.

Equation (3.20) is further used in a dimensionless form; for this purpose, we multiply both its parts by d_p/ν. As a result, we have:

$$\frac{V_x d_p}{\nu} = \frac{g\tau_{p0}d_p}{\nu\left[1 + \frac{1}{6}\left(\frac{V_x d_p}{\nu}\right)^{2/3}\right]}. \qquad (3.21)$$

The following expression derived from (3.21) may be used to estimate the precision of LDA measurements of the rates of sedimentation of particles:

$$Re_p = \frac{Ar_p}{18\left(1 + \frac{1}{6}Re_p^{2/3}\right)}, \qquad (3.22)$$

where $Re_p = V_x d_p/\nu$ is the Reynolds number of a particle, calculated by the particle velocity; and $Ar_p = g\rho_p d_p^3/\rho\nu^2$ is the Archimedes number for the particle. Relation (3.22) is based on the empirical (repeatedly verified and widely covered in the literature) dependence for the coefficient of aerodynamic drag for non-Stokesian particles.

Figure 3.12 gives the results of comparison of the measured (using the above-described techniques of optimization of LDA parameters) values of averaged velocities of polydisperse large particles with those predicted by relation (3.22). Some parameters of the transmitting and receiving systems of LDA 10 anemometer (Dantec) used in test experiments are given in Table 3.2. Based on these data, one can make an inference about good agreement between experiment and theory.

3.4.4 Measurement of the Relative Concentration of Particles

The concentration of particles is one of the basic physical parameters which define the characteristics of motion of the dispersed phase, as well as the

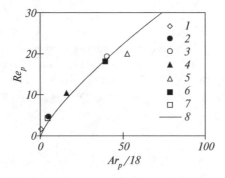

Fig. 3.12. Comparison of the calculated and measured velocities of large particles: (1) glass (50 µm), (2) glass (100 µm), (3) glass (200 µm), (4) iron (100 µm), (5) iron (150 µm), (6) copper (130 µm),(7) lead (59 µm), and (8) curve by relation (3.22)

Table 3.2. Basic parameters of the optical system of LDA

no.	parameter	value
1	laser radiation wavelength λ, µm	0.5145
2	laser beam diameter (level e^{-2})d_{L}, mm	1.15
3	beam expansion coefficient E	1.95
4	focal distance of the front lens f, mm	310
5	receiving optics angle α, deg.	7
6	beam intersection angle θ, deg.	1.98
7	dimensions of the measuring volume (level e^{-2}) $d_x \times d_y \times d_z$, mm	0.091×0.091×5.24
8	interference lattice spacing δ_{f}, µm	14.87
9	number of spacing in the measuring volume N_{f}	6

extent to which the dispersed phase affects the flow of the carrier medium. The distribution of particles in space may be significantly nonuniform, for example, in the vicinity of the surface of a body subjected to a heterogeneous flow. Because of this, correct measurements of the concentration of large particles present an urgent problem.

It is the objective of this section to review the capabilities of an LDA commercially produced by Dantec as regards the measurement of the distributions of relative concentration of large particles under conditions of a dust-laden gas flow [32].

Two methods of measuring the relative concentration of particles are described later along with the data of test experiments.

Conditions of Test Measurements

The experiments were performed for a downward turbulent flow of air in a pipe of inside diameter $D = 64$ mm. The Reynolds number was $Re_{\mathrm{D}} = 11,200$ with the averaged velocity of air on the pipe axis $U_{xc} = 2.8\,\mathrm{m\,s^{-1}}$. The distribution of the relative concentration of particles was measured in the cross-section spaced at distance $L = 1{,}280\,\mathrm{mm}$ from the beginning of the pipe and the point of inlet of the dispersed phase. The pipe average mass flow-rate concentration of particles was determined by the weighing method and was $\langle M_G \rangle = 0.4$. The true mass and mass flow-rate concentrations of particles are related by:

$$\frac{\langle M \rangle}{\langle M_G \rangle} = \frac{\langle U_x \rangle}{\langle V_x \rangle}, \tag{3.23}$$

where $\langle U_x \rangle$ and $\langle V_x \rangle$ denote the pipe cross-section average velocities of the carrier phase and particles, respectively.

For conditions of our experiments, $\langle M \rangle = 0.36$. The delivery of particles was organized in the cross-sectional region of the pipe so that the particle distribution in the measuring cross-section would be other than uniform.

The dispersed phase was provided by spherical glass particles of mass average diameter $d_p = 100\,\mu\mathrm{m}$.

Methods of Measuring the Particle Concentration

An LDA 10 two-channel three-beam laser Doppler anemometer (manufactured by Dantec, Denmark) operating according to a differential scheme was used in the experiments. Only one channel of this LDA was used in developing the procedure of measurement of the particle concentration and in performing the experiments. A Doppler signal processor of the counter type (55L90a LDA Counter Processor) was used for measuring the particle concentration. The basic parameters of the optical system of the LDA used in the experiments are given in Table 3.3.

Note that the investigations were performed with a view to measuring the relative concentration of particles for a pipe flow. The absolute cross-section average value of the particle concentration was determined by the weighing method. The determination of local values of absolute concentration of large particles is complicated by the fact that large particles scatter light from a volume whose magnitude exceeds significantly that of the LDA measuring volume determined by the intensity level e^{-2}. This was demonstrated earlier. A number of researchers tried to find the value of the measuring volume for particles of an arbitrary size and determine the absolute value of their concentration [2, 15, 24].

Measurement of the Relative Concentration of Particles by the Frequency of Incoming Signals

The data rate is an extensive quantity because it depends directly on the concentration of light-scattering particles in the measuring volume, on their velocity, and on the sensitivity of the receiving system of the LDA.

We will use simple relations to demonstrate the correlation between the particle concentration and the basic characteristics of Doppler signals. The

Table 3.3. Basic parameters of the optical system of LDA

no.	parameter	value
1	laser radiation wavelength λ, μm	0.5145
2	laser beam diameter (level e^{-2}) d_{L}, mm	1.15
3	beam expansion coefficient E	1.95
4	focal distance of the front lens f, mm	310
5	polarization of probing beams	vertical
6	aperture D of the receiving optical system, mm	47
7	diameter d of the diaphragm of the receiving optical system, mm	0.1
8	angle β between the axis of the receiving optical system and the plane normal to the probing plane, deg.	3
9	angle α between the axis of the receiving optical system and the probing plane, deg.	7
10	beam intersection angle θ, deg.	9.08
11	dimensions of the measuring volume (level e^{-2}) $d_x \times d_y \times d_z$, mm	0.091×0.091×1.15
12	interference lattice spacing δ_{f}, μm	3.25
13	number of spacings in the measuring volume N_{f}	28

total rate of data from the particles of the dispersed phase may be determined as:

$$D_{\mathrm{p}} = \frac{D_{\mathrm{pV}}}{V}, \tag{3.24}$$

where D_{pV} is the "reliable" data rate.

The total signal rate and the particle concentration in a flow are related as:

$$D_{\mathrm{p}} = N \times V_x \times S, \tag{3.25}$$

where N is the number density of particles (the number of particles per unit volume), V_x is the projection of the particle velocity on the direction of measurements, and S is the area of the section through the "effective" volume by a plane normal to the direction of measurements.

Most frequently employed in the theory of heterogeneous flows is the concept of mass concentration of particles, the expression for which has the following form in the case of low-volume content of the dispersed phase:

$$M = \frac{\Phi \rho_{\mathrm{p}}}{\rho} = \frac{N \rho_{\mathrm{p}} \pi d_{\mathrm{p}}^3}{6\rho}. \tag{3.26}$$

We use expressions (3.25) and (3.26) to find the correlation between the mass concentration of particles in a flow and the total data rate,

$$M = \frac{D_{\mathrm{p}} \rho_{\mathrm{p}} \pi d_{\mathrm{p}}^3}{6\rho V_x S}. \tag{3.27}$$

We write relation (3.27) for some arbitrary point of a heterogeneous flow (all parameters at this point will be indicated by the subscript 0). Under conditions in which ρ, ρ_p, and d_p are constants, relation (3.27) assumes the form:

$$M_0 = \frac{D_{p0}\rho_p \pi d_p^3}{6\rho V_{x0} S_0}. \tag{3.28}$$

Given the invariant sensitivity of the LDA (which is defined primarily by the value of voltage applied to the PM) and the same parameters of the optoelectronic system, we can assume that $S = S_0$. As a result, relations (3.27) and (3.28) yield the correlation between the values of mass concentration of particles at two points of flow,

$$\frac{M}{M_0} = \frac{D_p V_{x0}}{D_{p0} V_x}. \tag{3.29}$$

The thus obtained expression leads one to infer that the total signal rates and the velocities of light-scattering particles must be measured in order to determine the relative concentration of disperse inclusions.

Figure 3.13 gives the distribution of the mass concentration of particles over the pipe cross-section (r is the distance from the pipe axis), which was obtained as follows:

1. The data rates were measured at ten selected points over the pipe cross-section $D_{pV} = D_{pV}(r)$ at a constant sensitivity of the LDA (the voltage applied to the PM was $U = 1\,\text{kV}$). Simultaneously with the data rate, the reliability of data at the foregoing points was registered in order to derive the relation $V = V(r)$.
2. The total data rate was determined by relation (3.24), i.e., $D_p(r) = \frac{D_{pV}(r)}{V(r)}$.
3. The particle velocity distribution over the pipe cross-section was measured, $V_x = V_x(r)$.
4. Some function $\gamma = \gamma(r)$ was obtained by dividing $D_p = D_p(r)$ by the local value of velocity, i.e., $\gamma(r) = \frac{D_p(r)}{\bar{V}_x(r)}$, where $\bar{V}_x(r) = \frac{V_x(r)}{V_{xc}}$. Here, V_{xc} is the particle velocity on the pipe axis. According to relation (3.29), a

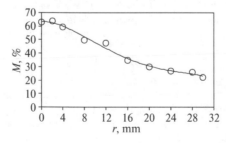

Fig. 3.13. The distribution of the mass concentration of particles over the pipe cross-section

local value of mass concentration of particles is directly proportional to the obtained function, i.e., $M \sim \gamma$.

5. The average (over the pipe cross-section) value of function γ was found by the relation

$$\langle \gamma \rangle = \frac{2}{R^2} \int_0^R \gamma(r) r \, dr.$$

6. The average (over the cross-section) value of mass concentration (which may be readily found given the values of $\langle M_\mathrm{G} \rangle$, $\langle U_x \rangle$, and $\langle V_x \rangle$) was placed in correspondence with the obtained value of $\langle \gamma \rangle$. The sought distribution of the particle concentration was determined as:

$$M(r) = \frac{\gamma(r)}{\langle \gamma \rangle} \langle M \rangle. \tag{3.30}$$

Figure 3.13 demonstrates clearly that the value of mass concentration of particles on the pipe axis is $M \approx 0.6$ and is three times the respective value in the wall region.

The estimates reveal that the procedure described earlier enables one to perform measurements of the relative concentration of particles with an error of 15% or less. Note that the main contribution to this error of measurement is made by the accuracy of determining the total data rate.

Measurement of the Relative Concentration of Particles by the PM Anode Current

A microammeter, which operates in the range of 0–100 μA and measures the strength of the anode current from the PM, is used to measure the relative concentration of particles by this method. The current strength depends on the LDA sensitivity and the physical and optical properties of light-scattering particles, as well as on the particle concentration and velocity. It is obvious that, for identical particles ($d_\mathrm{p} = $ const, $\rho_\mathrm{p} = $ const.) at a constant sensitivity of the LDA, the current strength will depend on the particle concentration and velocity, i.e., $I \sim M V_x$. However, significant difficulties are involved in deriving the theoretical dependence of the anode current $I = I(M, V_x)$. Therefore, the ammeter was calibrated directly by delivering a heterogeneous flow with known (previously measured) distributions of the particle concentration and velocity.

The sequence of evaluation of this method of determining the particle concentration was as follows:

1. Measurements of the current strength $I = I(r)$ at the same ten selected points over the pipe cross-section under the conditions described earlier for different values of voltage on the PM.
2. Measurements of the distribution of particle velocity over the pipe cross-section $V_x = V_x(r)$.

3. Obtaining the relation $I' = I'(r)$ by way of dividing $I = I(r)$ by the local value of velocity $\bar{V}_x = \bar{V}_x(r)$, i.e., $I'(r) = I(r)/\bar{V}_x(r)$. The obtained distributions $I' = I'(M)$ are given in Fig. 3.14. In so doing, the values of particle concentration were taken to be known and borrowed from Fig. 3.13. One can well see that the current strength for the conditions of our experiments is directly proportional to the value of mass concentration.

4. Determining the average (over the pipe cross-section) value of I' by the relation

$$\langle I' \rangle = \frac{2}{R^2} \int_0^R I'(r) r \mathrm{d}r.$$

5. The average (over the cross-section) value of mass concentration $\langle M \rangle$ was placed in correspondence with the obtained value of $\langle I' \rangle$. As a result, we have the distribution of particle concentration as:

$$M(r) = \frac{I'(r)}{\langle I' \rangle} \langle M \rangle. \tag{3.31}$$

Figure 3.15 gives the concentration of the dispersed phase, measured by the value of PM current, for different values of LDA sensitivity. This figure shows good agreement between the distributions of particle concentration, obtained using different methods.

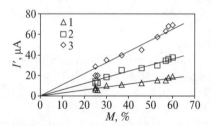

Fig. 3.14. The PM current as a function of mass concentration of particles: (1) $U = 0.85\,\mathrm{kV}$, (2) $U = 1\,\mathrm{kV}$, and (3) $U = 1.15\,\mathrm{kV}$

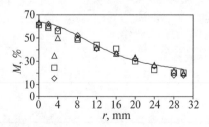

Fig. 3.15. The distribution of mass concentration of particles over the pipe cross-section: (1) $U = 0.85\,\mathrm{kV}$, (2) $U = 1\,\mathrm{kV}$, (3) $U = 1.15\,\mathrm{kV}$; the curve indicates the distribution of concentration obtained by the data rate

The inaccuracy of measurement of the PM current strength makes the main contribution to the error of measurement of the relative concentration of particles. The estimates have shown this error to be within 25%.

3.5 Special Features of Studies of the Effect of Solid Particles on Gas Flow

In order to study the inverse effect of particles on the characteristics of gas flow, it is necessary to perform measurements of the fields of instantaneous velocities of tracer particles, which simulate the motion of a continuous medium, in the presence of particles of the dispersed phase and subsequent statistical treatment of the measurement results. The main problem involved in performing such measurements in heterogeneous flows consists in the probability of emergence of cross-talk from particles of both types (small tracer particles and large particles of the dispersed phase) present in the flow. The accuracy of the results depends largely on the extent to which the signals may be separated from the foregoing particles. Note that the LDAs commercially produced by TSI (USA) and Dantec (Denmark) are not equipped with signal selection devices. Nevertheless, in recent years the LDA has become the basic tool employed in the investigation of heterogeneous flows. In performing such experiments, researchers were forced to develop devices for signal discrimination. Described later are methods of signal selection and the procedure of theoretical estimation of the efficiency of amplitude selection of signals and its experimental monitoring.

3.5.1 Estimation of Cross-Talk: Methods of Signal Selection

Prior to making provisions for signal selection, the expected cross-talk Ψ may be estimated as follows:

$$\Psi = \frac{D_{\mathrm{p}}''}{D_{\Sigma}''} = \frac{D_{\mathrm{p}}''}{D_{\mathrm{pf}}'' + D_{\mathrm{p}}''}, \tag{3.32}$$

where D_{p}'' is the rate of data from particles of the dispersed phase when performing measurements in a heterogeneous flow, D_{Σ}'' is the total data rate, and D_{pf}'' is the rate of data from tracer particles when performing measurements in a heterogeneous flow.

We will determine the rate of data from the particles of the dispersed phase when performing measurements in a heterogeneous flow by taking this rate to be equal to the number flow rate of solid particles through the cross-section of the measuring volume of the LDA, i.e., $D_{\mathrm{p}}'' = NV_x S$.

The number flow rate of particles of the dispersed phase through the measuring volume may be determined as:

$$NV_x S = \frac{G_{M\mathrm{p}}}{m_{\mathrm{p}}} \tag{3.33}$$

where G_{Mp} is the mass flow rate of particles of the dispersed phase through the cross-section of the measuring volume, and m_p is the mass of a single particle of the dispersed phase.

We will determine the mass flow rate of particles as:

$$G_{Mp} = M_G G_{Mg} = M_G \rho U_x S, \tag{3.34}$$

where M_G is the mass flow-rate concentration of the dispersed phase, G_{Mg} is the mass flow rate of the carrier phase through the cross-section of the measuring volume, ρ is the density of the carrier phase, U_x is the averaged velocity of the carrier phase in the direction of measurements, and S is the cross-sectional area of the measuring volume.

We take $S \approx d_x d_z$ and $m_p = \rho_p \frac{\pi d_p^3}{6}$ to derive

$$D_p'' \approx M_G U_x \left(\frac{6}{\pi d_p^3}\right) d_x d_z \left(\frac{\rho}{\rho_p}\right), \tag{3.35}$$

where d_x and d_z are the dimensions of the measuring volume of the LDA in the direction of measurements and along the optical axis, respectively.

The value of the total data rate in performing measurements in a heterogeneous flow may be determined with adequate accuracy only experimentally. For preliminary estimation, we will ignore the effect of the presence of solid particles on the rate of data from tracer particles and will take the total data rate to be equal to the characteristic value in the case of LDA measurements in single-phase flows, i.e., $D_\Sigma'' \approx 10^4$. As a result, the cross-talk will be defined by the expression:

$$\Psi \approx \frac{6\rho M_G U_x d_x d_z}{\rho_p \pi d_p^3 D_\Sigma''}. \tag{3.36}$$

Therefore, if the number density of particles of the dispersed phase is negligible compared to the number density of tracer particles (for example, in flows with relatively large particles at low values of the mass flow-rate concentration), the cross-talk will be weak. In this case, there is no need for signal selection, and the procedure of measurements in a heterogeneous flow will not differ from the procedure of standard LDA measurements. The foregoing is supported graphically by Fig. 3.16 which gives the values of cross-talk, calculated by formula (3.36), as a function of the mass concentration of particles of the dispersed phase and of their size. The estimates were made using the following values characteristic of LDA investigations of air flows with solid particles: $U_x = 10\,\mathrm{m\,s^{-1}}$, $d_x = 0.1\,\mathrm{mm}$, $d_z = 1.0\,\mathrm{mm}$, $\rho/\rho_p = 0.001$, and $D_\Sigma'' = 10^4$. We will take the cross-talk value of 1% to be its adequate value. Then, one can infer from Fig. 3.16 that the signal selection is necessary for particles of sizes $d_{p1} = 100\,\mu\mathrm{m}$, $d_{p2} = 200\,\mu\mathrm{m}$, $d_{p3} = 500\,\mu\mathrm{m}$, and $d_{p4} = 1{,}000\,\mu\mathrm{m}$, starting with the following values of mass concentration: $M_{G1} = 1\%$, $M_{G2} = 5\%$, $M_{G3} = 100\%$, and $M_{G4} = 1{,}000\%$.

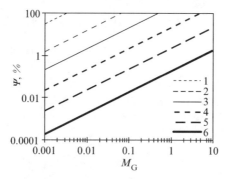

Fig. 3.16. The cross-talk as a function of mass concentration and particle size:
(1) 20 μm, (2) 50 μm, (3) 100 μm, (4) 200 μm, (5) 500 μm, and (6) 1,000 μm

Basic methods of signal selection of the following three types exist [3, 13]:

1. The method based on the intensity of radiation scattered by particles. The method is further based on the fact that the amplitude of a signal from large particles of the dispersed phase is higher than that from tracer particles defining the gas phase. The effective signal extraction may be accomplished by suppressing the signal of higher amplitude as overload.
2. The method based on the dependence of the visibility of a Doppler signal on the size of light-scattering particles. The use of this method presumes simultaneous measurements of particle velocities and sizes.
3. The method based on the Doppler signal frequency. The use of this method is possible in the presence of a difference between the instantaneous velocities of carrier gas and disperse particles, i.e., when the probability density curves for the velocities of both phases do not intersect. In this case, one can readily eliminate the cross-talk by operating in the frequency range in which the signals of the phase of interest to us are located.

The most efficient and simple to realize of the methods of signal selection described earlier in the case of using commercially available LDAs is the method of amplitude discrimination. As was already noted, a signal from large particles of the dispersed phase is higher in amplitude than that from tracer particles. Therefore, by lowering the threshold amplitude for the input signal admitted for further processing, one can significantly reduce the rate of data from particles of the dispersed phase and, consequently, the value of cross-talk. It does not appear possible to fully eliminate the cross-talk, because large particles are also capable of producing a signal of low amplitude (lower than the operation threshold of the discriminator) when they pass in the vicinity of the external boundary of the measuring volume (up to a distance of $nd_z/2$ from the axis of the measuring volume $(n > 1)$, where the amplitude is equal to the minimal threshold of operation (sensitivity) of the optoelectronic system (see Fig. 3.17)). This fact restricts the use of the present method of signal

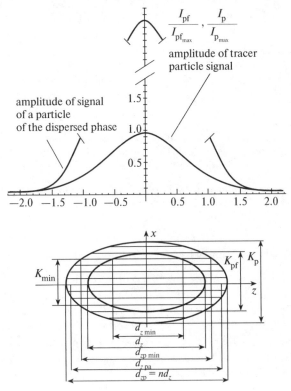

Fig. 3.17. The distribution of the intensity of scattered light for large and small particles and the respective measuring volumes

selection as regards the particle size and concentration. For example, Modarress et al. [14] observed that the amplitude method performs adequately in the presence of particles more than $200\,\mu m$ in size in the flow with a mass flow-rate concentration $M_G = 0.8$.

3.5.2 Estimation of the Efficiency of Amplitude Selection of Signals

This section contains a description of the procedure for approximate estimation of the efficiency of amplitude selection of signals from particles of the dispersed phase when performing measurements of the velocity of the gas phase of heterogeneous flow.

We will write the following expression for the amplitude of signal from a light-scattering tracer particle as a function of the distance z to the axis of the measuring volume [2] (see Fig. 3.17):

$$I_{\mathrm{pf}}(z) = I_{\mathrm{pf\,max}} \exp\left[-2\left(\frac{z}{d_z/2}\right)^2\right], \qquad (3.37)$$

$$I_{\mathrm{pf}}(0) = I_{\mathrm{pf\,max}} = \frac{8P_L e\eta_q\eta_{\mathrm{pf}}G(d_{\mathrm{pf}})}{\pi d_f^2 k^2 hc_L}, \tag{3.38}$$

Here, $I_{\mathrm{pf}}(0)$ is the amplitude of signal from a tracer particle passing through the center of the measuring volume, P_L is the laser radiation power, e is the elementary charge, η_q is the quantum efficiency, η_{pf} is the visibility of the tracer particle signal, $G(d_{\mathrm{pf}})$ is the dissipative function, d_f is the diameter of focused laser beam at the waist, k is the wave number, h is the Planck constant, and c_L is the velocity of light.

We will define the amplitude of signal of a tracer particle on the boundary of the measuring volume (determined by the intensity level e^{-2}) $I_{\mathrm{pf}}(d_z/2) = I_{\mathrm{pf\,max}}\exp(-2)$ as the threshold of sensitivity of the receiving optoelectronic system.

The following expressions similar to (3.37) and (3.38) may be written for the amplitude of signal from a large particle of the dispersed phase $(d_{\mathrm{p}} > d_{\mathrm{pf}})$:

$$I_{\mathrm{p}}(z) = I_{\mathrm{p\,max}}\exp\left[-2\left(\frac{z}{d_z/2}\right)^2\right], \tag{3.39}$$

$$I_{\mathrm{p}}(0) = I_{\mathrm{p\,max}} = \frac{8P_L e\eta_q\eta_{\mathrm{p}}G(d_{\mathrm{p}})}{\pi d_f^2 k^2 hc_L}, \tag{3.40}$$

where $I_{\mathrm{p}}(0)$ is the amplitude of signal from a particle of the dispersed phase, which passes through the center of the measuring volume; η_{p} is the visibility of signal of the particle of the dispersed phase; and $G(d_{\mathrm{p}})$ is the dissipative function.

We use relations (3.37) and (3.39) to find the value of the "effective" measuring volume along the optical axis for particles of the dispersed phase d_{zp} (see Fig. 3.17). The problem consists in searching for some distance $z_{\mathrm{p}} = d_{\mathrm{zp}}/2 = nd_z/2$ in the case of which the amplitude of signal from large particles of the dispersed phase will be equal to the threshold of sensitivity of the receiving optoelectronic system,

$$I_{\mathrm{p}}(z_{\mathrm{p}}) = I_{\mathrm{p}}(nd_z/2) = I_{\mathrm{p\,max}}\exp(-2n^2) = I_{\mathrm{pf\,max}}\exp(-2). \tag{3.41}$$

Simple transformations of (3.41) produce:

$$n = \left(1 - \frac{1}{2}\ln\frac{I_{\mathrm{pf\,max}}}{I_{\mathrm{p\,max}}}\right)^{1/2}. \tag{3.42}$$

Therefore, the "effective" measuring volume for large particles of the dispersed phase along the optical axis will be:

$$d_{\mathrm{zp}} = nd_z = d_z\left(1 - \frac{1}{2}\ln\frac{I_{\mathrm{pf\,max}}}{I_{\mathrm{p\,max}}}\right)^{1/2}. \tag{3.43}$$

The amplitude selection of signals makes it possible to reduce the rate of delivery of data from particles of the dispersed phase D_{p}'' when performing measurements in a heterogeneous flow and thereby reduce the value of cross-talk. In order to simplify subsequent analysis, we will assume that the presence of tracer particles has no effect on the rate of delivery of data from particles of the dispersed phase, i.e., $D_p'' \approx D_p'$ (D_p' is the rate of delivery of data from particles of the dispersed phase in the absence of tracer particles). In this case, we will determine the effectiveness of amplitude discrimination (degree of suppression of signals from particles of the dispersed phase) from the expression

$$E_a = \frac{D_{pa}'}{D_p'}, \qquad (3.44)$$

where D_{pa}' is the "suppressed" rate of data from particles of the dispersed phase in the course of amplitude selection of signals.

In using (3.44), one can readily see that the degree of suppression of signals from the dispersed phase will be complete (equal to unity) in the case of $D_{pa}' = D_p'$. We will take the distribution (number density) of particles of the dispersed phase over the cross-section of the "effective" measuring volume to be uniform. Clearly, the rate of delivery of data D_p' is proportional to the measuring volume d_{zp}.

By reducing the amplitude of the input signal to be treated, we reduce the measuring volume region for large particles and, consequently, the rate of delivery of data. We find some distance z_{pa} in the case of which the value of the amplitude of signal from a particle of the dispersed phase will be equal to that of the maximal amplitude of signal from a tracer particle $I_{pf\,max}$,

$$I_p(z_{pa}) = I_{p\,max} \exp\left[-2\left(\frac{z_{pa}}{d_z/2}\right)^2\right] = I_{pf\,max}. \qquad (3.45)$$

It is obvious that the maximal effectiveness of signal selection may be attained only if the amplitude threshold of input signal is lowered to a value equal to that of the maximal amplitude of signal from a tracer particle (no reduction of the rate of delivery of data from tracer particles $D_{pf}'(D_{pf}'')$ occurs in this case). After simple transformations of (3.45), we have:

$$z_{pa} = \frac{d_z}{2}\left(-\frac{1}{2}\ln\frac{I_{pf\,max}}{I_{p\,max}}\right)^{1/2} \qquad (3.46)$$

or

$$d_{zpa} = 2z_{pa} = d_z\left(-\frac{1}{2}\ln\frac{I_{pf\,max}}{I_{p\,max}}\right)^{1/2}. \qquad (3.47)$$

We use (3.43), (3.44), and (3.47) to write

$$\frac{D_{pa}'}{D_p'} = \frac{d_{zpa}}{d_{zp}} = \frac{d_z\left(-\frac{1}{2}\ln\frac{I_{pf\,max}}{I_{p\,max}}\right)^{1/2}}{d_z\left(1-\frac{1}{2}\ln\frac{I_{pf\,max}}{I_{p\,max}}\right)^{1/2}}. \qquad (3.48)$$

Therefore, we have the following equation for the effectiveness of amplitude selection of signals:

$$E_a = \frac{D'_{pa}}{D'_p} = \left[1 - \frac{2}{\ln(I_{pf\,max}/I_{p\,max})}\right]^{-1/2}. \tag{3.49}$$

We will now analyze (3.49) for two extreme cases, namely:

1. When the size of particles of the dispersed phase is comparable to the size of tracer particles

$$\left(\frac{d_{pf}}{d_p} \to 1, \frac{I_{pf\,max}}{I_{p\,max}} \to 1\right) \text{ and } E_a = \frac{D'_{pa}}{D'_p} \to 0,$$

the amplitude selection of signals is not effective.
2. When the size of particles of the dispersed phase is much larger than the size of tracer particles

$$\left(\frac{d_{pf}}{d_p} \to 0, \frac{I_{pf\,max}}{I_{p\,max}} \to 0\right) \text{ and } E_a = \frac{D'_{pa}}{D'_p} \to 1,$$

the amplitude selection of signals will be effective (the degree of suppression of signals from particles of the dispersed phase will be complete).

For qualitative estimation of the effectiveness of amplitude selection of signals using relation (3.49), one needs to know the ratio between the maximal amplitudes of signals of tracer particles and particles of the dispersed phase; however, this cannot be done using expressions (3.37) and (3.39), because each one of these expressions contains two unknowns, namely, the signal visibilities η_{pf} and η_p and the dissipative functions $G(d_{pf})$ and $G(d_p)$.

As for all other quantities appearing in (3.37) and (3.39), they may be taken to be constant for particles of both types with the given adjustment of the optoelectronic system.

We can make the following assumptions in order to determine the ratio of signal amplitudes:

1. The signal visibility is inversely proportional to the diameter of light- scattering particle,

$$\frac{\eta_p}{\eta_{pf}} = \frac{d_{pf}}{d_p}. \tag{3.50}$$

2. We will consider three possible options with regard to unknown dissipative functions, namely,

$$\left[\frac{G(d_p)}{G(d_{pf})}\right]_i = \left(\frac{d_p}{d_{pf}}\right)^j, \tag{3.51}$$

where $i = 1, 2, 3$ and $j = 1.5, 2, 3$.

With the assumptions made, we derive the following versions of the ratio between the amplitudes of signals of particles of the dispersed phase and of tracer particles:

$$\left(\frac{I_{\mathrm{p\,max}}}{I_{\mathrm{pf\,max}}}\right)_i = \left(\frac{d_{\mathrm{p}}}{d_{\mathrm{pf}}}\right)^j , \tag{3.52}$$

where $i = 1, 2, 3$ and $j = 0.5, 1, 2$.

As a result, (3.49) yields three different relationships for the estimation of the effectiveness of amplitude selection of signals,

$$E_{\mathrm{a}_i} = \left[1 - \frac{2}{\ln(d_{\mathrm{pf}}/d_{\mathrm{p}})^j}\right]^{-0.5} , \tag{3.53}$$

where $i = 1, 2, 3$ and $j = 0.5, 1, 2$.

The effectiveness of the amplitude selection of signals as a function of the ratio between the diameters of particles of the dispersed phase and of tracer particles is given in Fig. 3.18. This figure supports the foregoing observation that the use of amplitude discrimination does not enable one to fully eliminate cross-talk.

A further increase in the effectiveness of signal selection using an amplitude discriminator may be attained by increasing the number of interference fringes required for monitoring the reliability of signal. This possibility was mentioned, for example by Rogers and Eaton [19] who measured the kinematic parameters of carrier air in the presence of copper particles 70 μm with mass concentration $M = 0.2$.

When an LDA 10 two-channel three-beam laser Doppler anemometer (manufactured by Dantec, Denmark) and a Doppler signal processor of the counter type (55L90a LDA Counter Processor) are employed, regarded as reliable is the signal (in operation in the "5/8" mode) from particles for which the difference between velocities determined by the time of flight of five and eight interference fringes is within the preassigned error. Signals accepted

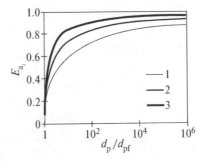

Fig. 3.18. The effectiveness of the amplitude selection of signals as a function of the ratio between the diameters of particles of the dispersed phase and of tracer particles (curves 1–3 are plotted by relations (3.53))

for statistical treatment are those from particles which passed not less than eight interference fringes of the measuring volume. If a large particle of the dispersed phase passes the boundary of the measuring volume and crosses a smaller number of interference fringes than that required for checking the reliability, the signal from this particle is not reliable and will not be accepted for further treatment. This factor causes a reduction of the "effective" measuring volume of particles of the dispersed phase. One must bear in mind that the requirement defining the minimal number of interference fringes leads to a reduction of the measuring volume for tracer particles as well (see Fig. 3.17); this causes a decrease in the rate of delivery of data from the latter particles and, therefore, is a factor causing in increase in cross-talk. It is only by reducing the "effective" measuring volume with the requirement defining the minimal number of interference fringes (Fig. 3.17) that a complete suppression of signals from particles of the dispersed phase may be accomplished $(D'_p(D''_p) = 0, \Psi = 0)$ provided the condition $d_{zpa} \geq d_{zp\,min}$ is valid.

In this case, the efficiency of the amplitude discriminator in performing measurements of the velocity of the gas phase of heterogeneous flow will be defined by the expression:

$$E_a = \frac{D'_{pa} + D'_{p\,min}}{D'_p}. \tag{3.54}$$

In view of the fact that the rates of delivery of data D'_p, D'_{pa}, and $D'_{p\,min}$, are proportional (in a first approximation) to d_{zp}, d_{zpa}, and $(d_{zp} - d_{zp\,min})$, respectively, we can write:

$$E_a = \frac{d_{zpa} + d_{zp} - d_{zp\,min}}{d_{zp}} \quad (0 \leq E_a \leq 1) \tag{3.55}$$

or, in view of (3.53),

$$E_{a_i} = 1 + \left[1 - \frac{2}{\ln(d_{pf}/d_p)^j}\right]^{-0.5} - d_{zp\,min}/d_{zp} \ (0 \leq E_{a_i} \leq 1), \tag{3.56}$$

where the ratio $d_{zp\,min}/d_{zp}$ is determined from (3.43) and (3.52) given the number of interference fringes required for checking the reliability of the signal and the parameters of the optical scheme (laser radiation wavelength, angle of intersection of laser beams forming the measuring volume, and the number of interference fringes in the measuring volume).

As a result, the region of possible values of cross-talk prior to experiments in measuring the parameters of the carrier phase of heterogeneous flow involving the use of amplitude selection of signals may be estimated as:

$$\Psi_{a_i} \approx (1 - E_{a_i})\Psi, \tag{3.57}$$

where the signal selection efficiency E_a is determined from (3.53) or (3.56), and the "initial" (without signal selection) value of cross-talk Ψ is determined from (3.36).

Monitoring of the Signal Selection

This section contains a description of the procedure for monitoring the efficiency of amplitude selection of signals when performing measurements of the parameters of the carrier phase.

In the experiments, the value of interference must be monitored, strictly speaking, at every measurement point. This may be done as follows. According to (3.32), the cross-talk is defined by the following expression:

$$\Psi = \frac{D_p''}{D_\Sigma''} = \frac{D_p''}{D_{pf}'' + D_p''}.$$ (3.58)

The value of D_Σ'' must be determined experimentally. We will use the foregoing assumption that $D_p'' \approx D_p'$ to determine the maximum possible value of cross-talk. The rate of delivery of data from particles of the dispersed phase in the absence of tracer particles D_p' may likewise be determined experimentally. As a result, the maximum expected cross-talk is found from the relation:

$$\Psi_{max} \approx \frac{D_p'}{D_\Sigma''}.$$ (3.59)

An example of experimental determination of cross-talk is given in Fig. 3.19. The basic parameters of flow are as follows: a rising turbulent flow of air in a pipe (channel axis) with glass particles 100 μm in diameter, the averaged velocity of air $U_x = 6.4\,\mathrm{ms}^{-1}$, $M = 0.26$, and the size of tracer particles (glycerin + water) $d_{pf} = 2\text{–}5\,\mu\mathrm{m}$.

The sequence of operations in determining the cross-talk was as follows:

1. Delivery of tracer particles into the flow.
2. Determination of the rate of delivery of data from tracer particles in a single-phase flow $D_\Sigma' = D_\Sigma'(I_e)$ with a varying amplitude threshold for the input signal from photomultiplier I_e given a constant sensitivity of the LDA optoelectronic system.

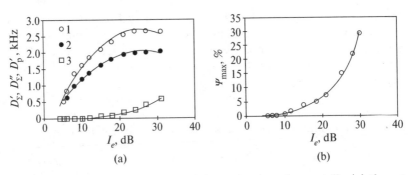

Fig. 3.19. An example of experimental determination of cross-talk: **(a)** the rate of delivery of data as a function of threshold voltage for the input signal ((1) D_Σ', (2) D_Σ'', (3) D_p'); **(b)** the maximal cross-talk as a function of threshold voltage

3. Termination of the delivery of tracer particles.
4. Delivery of particles of the dispersed phase.
5. Determination of the rate of delivery of data from particles of the dispersed phase $D'_{\mathrm{p}} = D'_{\mathrm{p}}(I_e)$ with a varying amplitude threshold for the input signal from photomultiplier I_e given a constant sensitivity of the LDA optoelectronic system. In so doing, the sensitivity (the voltage applied to the photomultiplier) must be equal to that observed when deriving the relationship $D'_{\Sigma} = D'_{\Sigma}(I_e)$.
6. Delivery of tracer particles into a heterogeneous flow.
7. Determination of the total rate of delivery of data $D''_{\Sigma} = D''_{\Sigma}(I_e)$ given a constant sensitivity of LDA.
8. Determination of the maximum possible cross-talk,

$$\Psi_{\max}(I_e) \approx \frac{D'_{\mathrm{p}}(I_e)}{D''_{\Sigma}(I_e)}, \tag{3.60}$$

Figure 3.19b demonstrates that, in the case of threshold voltage $I_e = 10\,\mathrm{dB}$, the value of cross-talk does not exceed 1%, which may be regarded as quite acceptable in performing measurements of the velocity of the carrier phase of a heterogeneous flow.

3.6 Experimental Apparatuses

The wealth of experience gained from investigations of turbulent single-phase flows is used in designing experimental apparatuses to study turbulent heterogeneous flows. Nevertheless, some concrete problems may be identified, which the researchers of particle-laden flows must solve. These problems include the introduction of the dispersed phase into a turbulent flow of gas, development of conditions for hydrodynamic and thermal stabilization of heterogeneous flow, and the trapping of particles.

Two experimental facilities have been developed at the Institute of High Temperatures of the Russian Academy of Sciences (IVTAN) for the investigation of the structure of turbulent heterogeneous flows, those for studying upward and downward flows of gas suspension. Both facilities operate in an open (with respect to gas and dispersed phase) circuit. This scheme enables one to determine the concentration of solid particles with the desired accuracy by way of direct separate measurements of the flow rate of gas and solid particles. The use of an open circuit further enables one to fairly simply adjust the flow rates of both phases of heterogeneous flow. In the overwhelming majority of studies of the structure, hydraulic drag, and heat transfer of dust-laden flows, air was used as the carrier medium. This is explained by its availability and low cost compared to other gases, which is very important in the case of open-circuit facilities. Note that the use of air is further convenient from considerations of the necessity of comparing the measurement results with the data of other researchers for generalizing the results.

3.6.1 Experimental Setup for Studying Upward Flows of Gas Suspension

A basic diagram of the setup is given in Fig. 3.20. Air is delivered via silencer into the inlet pipe of an RGN-300 gas blower with the maximal capacity of $0.5\,\mathrm{kg\,s^{-1}}$ at excess pressure of $0.03\,\mathrm{MPa}$. The flow rate is adjusted by way of by-passing air to the atmosphere using by-pass and mainline gates remote-controlled from the operator's desk. The flow rate is measured by a flowmeter in the form of a precalibrated "quadrant" nozzle. Air is then passed into a receiver required for damping pulsations from the gas blower, passes through a segment of the delivery line, and comes to the inlet of the experimental section. This inlet consists of a rectangular tank with an inner frame for increasing the strength and rigidity of the structure. The tank has a flange on its upper face, to which a sleeve and a vertical pipe of 12Kh18N10T

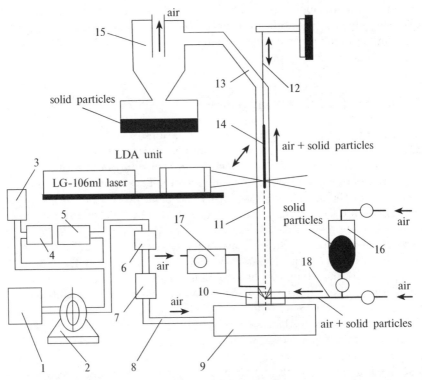

Fig. 3.20. Schematic of the experimental setup for studying the structure of upward heterogeneous flows: (1) silencer, (2) gas blower, (3) by-pass gate, (4) mainline gate, (5) by-pass line, (6) flowmeter, (7) receiver, (8) segment of delivery line, (9) inlet of the experimental section, (10) sleeve, (11) vertical pipe, (12) bar, (13) reducer, (14) rod with a hemispherical end, (15) cyclone, (16) solid particle delivery unit, (17) generator of micrometer-sized tracer particles, (18) conveying air

(chrome–nickel–titanium) stainless steel with an inside diameter of 64 mm and wall thickness of 6 mm are secured. The pipe is 1,500 mm long. A slit for the inlet and outlet of LDA probing beams (LDA 10 by Dantec) is provided in the pipe wall at a distance of 300 mm from the upper end. In order to provide for tightness of the experimental section, the slit is covered by transparent windows attached to the pipe by tie pins such that the planes of the windows are located normally to the optical axis of the LDA transmitting optics. A rod serving as a body subjected to flow is placed coaxially in the pipe by means of a bar and a guide tube welded into a reducer. The flow passes the experimental section and the reducer to get into a TsN15Y cyclone, after which it is discharged into the atmosphere.

A displacement-type unit for the delivery of solid particles was used to develop a heterogeneous flow. The operating principle of this delivery unit is as follows. Solid particles are poured into a 3-l glass bottle. When compressed air is fed into the bottle, the particles are forced out into the delivery line where they are picked up by conveying air. The resultant dust–air mixture is delivered via distributing union into the circuit of the experimental section. This structure of delivery unit makes possible a wide-range variation of the flow rate of particles depending on the pressure difference across the delivery unit and provides for the time constancy of the flow rate.

A 55L18 generator of micrometer-sized particles (made by Dantec, Denmark) utilizing a glycerin–water mixture is used to introduce tracer particles into the flow, which simulate the motion of carrier medium. Particles 2–5 μm in size are generated.

The measuring region is scanned using two 57H00 traverse systems by Dantec which make it possible to automatically move a rod with an accuracy of ±10 μm along the vertical axis and to move the measurement point proper (LDA measuring volume) with the same accuracy along the horizontal axis.

3.6.2 Experimental Setup for Studying Downward Flows of Gas Suspension

A diagram of the setup is given in Fig. 3.21. The working section is a vertical pipe of 12Kh18N10T (chrome-nickel-titanium) stainless steel with an inside diameter of 46 mm and wall thickness of 2 mm. The pipe is 2,500 mm long. A slit 12 mm wide for the inlet and outlet of probing beams of the laser Doppler anemometer (LDA 10 by Dantec) is provided in the pipe wall at distance $L = 1,380$ mm from the upper end. Air is delivered into the working section via receiver from bottles into which it is prepumped by a compressor (model K2-150). In order to develop a heterogeneous flow, particles are charged into a 2-l "feeder" and poured downwards via opening in the "feeder" cover and a centering tube welded into the reducer. The flow rate of particles is varied by using "feeder" covers with openings of different diameters. The particles are taken up by the air flow, pass the working section, and are deposited in a gravitational chamber.

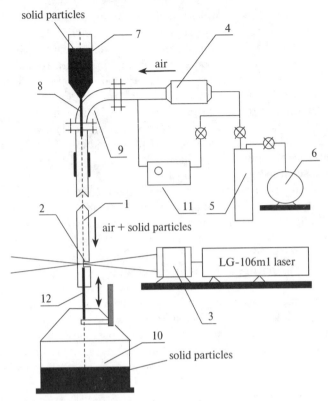

Fig. 3.21. Schematic of the experimental setup for studying the structure of downward heterogeneous flows: (1) vertical pipe, (2) slit for the inlet and outlet of LDA beams, (3) unit of LDA transmitting optics, (4) receiver, (5) bottles with compressed air, (6) compressor, (7) solid particle feeder, (8) centering tube, (9) redicer. (10) gravitational precipitation chamber, (11) generator of micrometer-sized tracer particles, (12) rod with a hemispherical end

Micrometer-sized tracer particles are introduced into the flow for performing measurements of the carrier phase velocity; the Dantec-made generator of particles is used for this purpose. The pipe cross-section is scanned using a traverse system which makes it possible to automatically move the measuring volume with an accuracy of $\pm 10\,\mu$m.

3.6.3 The Choice of Particle Characteristics: An Example

The physical properties of particles (first of all, their size and density) and the working range of concentrations must be selected strictly in accordance with the objectives of investigation. For example, it was predominantly non-equilibrium flows that served the subject of experimental investigations at the IVTAN facilities described earlier (see Table 1.1). Heterogeneous flows

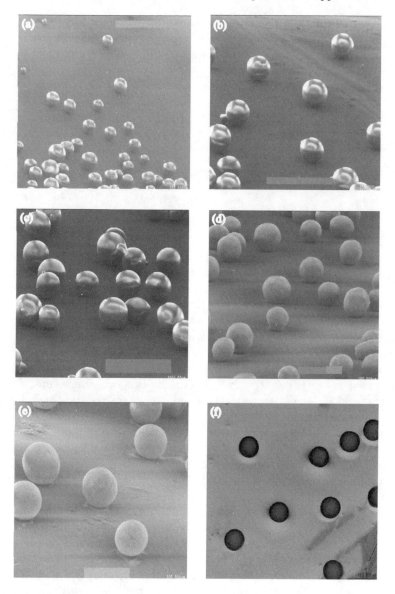

Fig. 3.22. Photographs of particles of: (**a**) glass (50 μm), (**b**) glass (100 μm), (**c**) glass (200 μm), (**d**) iron (100 μm), (**e**) iron (150 μm), and (**f**) lead (59 μm)

of this class are characterized by the presence of interphase dynamic slip in averaged and fluctuation motions. The Stokes numbers in averaged and large-scale fluctuation motions for such flows are $Stk_f = \tau_p/T_f \approx O(1)$ and $Stk_L = \tau_p/T_L \approx O(1)$. The characteristic time of the carrier phase in average

Table 3.4. Parameters of employed spherical particles

no.	material	rated diameter, μm	mean-square deviation of the diameter, μm	density of the particle material, $kg\,m^{-3}$	time of dynamic relaxation (Stokes law), s	country
1	SiO_2	50	5	2,550	0.020	Japan
2	SiO_2	100	8	2,550	0.08	Japan
3	SiO_2	200	16	2,550	0.32	Japan
4	Fe	100	9	7,800	0.24	Japan
5	Fe	150	12	7,800	0.54	Japan
6	Cu	130	18	8,900	0.46	Russia
7	Pb	59	0.6	11,340	0.12	Russia

motion T_f usually exceeds the integral Lagrangian scale of turbulence T_L which characterizes the lifetime of energy-carrying eddies. For obtaining a steady-state heterogeneous flow (ensuring a complete acceleration of particles) at the measuring cross-section, it is necessary that the characteristic time of carrier gas would be at least several times that of dynamic relaxation of particles τ_p. The characteristic time of carrier gas in averaged motion may be estimated as the ratio of the length of experimental section (distance from the point of injection of particles to the measuring cross section) to the characteristic value of averaged velocity, i.e., $T_f = L/U_x$. For the experimental setups, $L = 1.2$ to 1.4 m, and $U_x = O\,(10\,\text{m/s})$. Therefore, the characteristic time of carrier gas in averaged motion T_f amounts to hundreds of milliseconds. Because of this, the time of relaxation of particles $\tau_p \approx \rho_p d_p^2/18\,\mu$ must be of the order of 0.1 s or less. The size and density of particles of the dispersed phase were selected in view of this requirement. Photographs of some particles used in the experiments are given in Fig. 3.22. The basic parameters of solid particles are given in Table 3.4.

The subject of investigations was largely provided by weakly dust-laden flows in which the presence of the dispersed phase has an inverse effect on the characteristics of carrier gas. The volume concentration of particles in such flows varies in the range $\Phi = 10^{-3}$–10^{-6} (Fig. 1.8). The presence of solid particles (with a relative density of the material $\rho_p/\rho = 10^3$–10^4) predetermines the following working range of values of mass concentration: 0.001–$0.01 < M < 1$–10.

4

Particle-Laden Channel Flows

4.1 Preliminary Remarks

Studying the behavior of solid particles in a turbulent flow and their inverse effect on the characteristics of carrier gas presents one of the fundamental problems in the mechanics of heterogeneous media. The features characteristic of particle motion and the intensity of interphase processes are largely defined by the inertia of the dispersed phase and its concentration in the flow.

The investigation of heterogeneous flows in channels (in particular, in pipes) is not a trivial problem. Studying the motion of particles in the flow field of the gas carrying them in the presence of gradients of averaged and fluctuation velocities and temperatures (in the case of nonisothermal flow) in the radial direction is not a simple problem per se. The gradient pattern of the profiles of averaged and fluctuation parameters of carrier gas leads to the nonuniformity of the force factors acting on a particle in the longitudinal and radial directions. This causes the formation of significantly nonuniform profiles of averaged and fluctuation velocities, temperatures, and concentrations of particles. The presence of shear profiles of characteristics of particles makes rather difficult the study of their inverse effect on the characteristics of the carrier medium. Therefore, as a result of their complexity, heterogeneous flows in pipes remain little-studied in spite of the numerous investigations of these flows.

In this chapter, we will describe and analyze the results of investigations of turbulent flows with solid particles in channels. Section 4.2 is devoted to treatment of the characteristics of motion of particles and carrier phase under conditions of heterogeneous flows of different types. In Sect. 4.3, the model is described which was developed for the inclusion of the effect of the dispersed phase on the turbulent energy of gas.

4.2 The Behavior of Solid Particles and Their Effect on Gas Flow

In this section, we will treat the experimental data on the distribution of averaged and fluctuation velocities of a "pure" gas, of a gas in the presence of particles, and of solid particles proper for heterogeneous flows in channels under conditions of the concentration and inertia of the dispersed phase varying in a wide range. Also described below are various attempts of generalizing the available experimental data.

Historically, it turned out that the theory of turbulent heterogeneous jets made greater progress in its development. This is associated with the obviousness of the practical application of dust-laden flows of this class and with the relative simplicity of performing experimental investigations. Early studies of heterogeneous jets [15, 20] revealed that the presence of particles leads to a decrease in the intensity of turbulence of carrier gas. The particles caused a variation in the energy spectrum of turbulence as well, by suppressing the high-frequency components. These results were confirmed by more recent investigations (for example, [8, 23, 30, 37] and others) performed for a wide range of particle concentrations and sizes, as well as of the density ratio of the phases.

Early investigations of heterogeneous flows in channels are also described in [4, 7, 26, 32–34]. They are largely devoted to flows with spherical particles in vertical pipes, with a mass flow rate concentration $M_G \leq 5$. A detailed review of papers published before 1969 is found in [26]. A typical tool employed in these experiments was provided by Pitot tubes for measuring the velocity of carrier gas and various photographic techniques for measuring the particle velocity. The investigations revealed the effect of particles on the profile of averaged velocity of the carrier phase in the case of mass concentration $M_G > 1$. It was impossible to directly measure the intensity of gas turbulence. Soo et al. [33] investigated the behavior of gas turbulence in the presence of particles using the diffusion of gas indicator. These measurements (valid in the vicinity of the pipe axis) failed to reveal any effect of particles on the intensity of turbulence. Nevertheless, a decrease in the integral Lagrangian scale of turbulence was demonstrated.

More recent investigations of heterogeneous flows in channels were largely performed using various modifications of LDA which enable one to measure the velocities of both phases [16, 19, 21, 24, 25, 29, 35, 36, 38, 39, 42–44].

4.2.1 Averaged Velocities of Gas and Particles

We will now turn to analyzing the presently available data on the distribution of averaged velocities of gas in the presence of particles and of the particles proper under conditions of heterogeneous flows of different types.

Equilibrium Flow

The case of equilibrium heterogeneous flow is an extreme case whose mathematical and physical simulation presents no serious difficulties. Low-inertia particles which are present in an equilibrium flow follow completely the turbulent fluctuations of the carrier gas velocity (see Table 1.1). As a result, the profile of averaged velocity of these particles will exactly repeat the respective profile for the carrier phase. In particular, such low-inertia particles are involved in the investigation of single-phase flows using LDA. The thus measured velocity of tracer particles is associated with the gas velocity. True, one must bear in mind the fact that the volume (mass) concentration of low-inertia tracer particles introduced into a single-phase flow is negligible. Because of this, the inverse effect on the characteristics of turbulence of the carrier phase (in particular, on the distribution of averaged velocity) is likewise insignificant. The effect of such particles on gas must increase with their concentration. I am not aware of experimental investigations aimed at studying equilibrium heterogeneous flows with significant concentrations of particles. However, Boothroyd [3] assumed that the increase in the concentration of such particles will bring about a variation of the physical properties of heterogeneous medium compared to those of a flow of "pure" gas. The density of material of solid (liquid) particles exceeds significantly the density of carrier gas, i.e., $\bar{\rho} = \rho_p/\rho \geq 10^3$; therefore, the "effective" density of heterogeneous flow must increase with the concentration of the dispersed phase. In a first approximation (ignoring the volume taken up by particles), this characteristic will be determined as follows:

$$\rho_e = \rho + \Phi\rho_p = \rho(1 + M). \tag{4.1}$$

The increase in the density of the carrier phase will result in a decrease in its kinematic viscosity,

$$\nu_e = \frac{\mu}{\rho_e}. \tag{4.2}$$

In turn, the decrease in the viscosity of gas will cause a variation of the basic process characteristic of turbulent flow, namely, the Reynolds number, the expression for which will take the form

$$Re_D = \frac{\langle U_x \rangle 2R}{\nu_e}. \tag{4.3}$$

One can infer from (4.3) that the presence of such low-inertia particles will cause in increase in the Reynolds number.

Quasiequilibrium Flow

Heterogeneous flows of this type are characterized by the equality of averaged velocities of the carrier and dispersed phases (see Table 1.1). The respective distributions of averaged velocities over the channel cross-section will also have a similar form. However, unlike the case of equilibrium flow, the particle inertia will be sufficient for the presence of difference between the fluctuation velocities of gas and suspended particles. Because the values of the Stokes number of these particles in large-scale fluctuation motion are of the order of unity, i.e., $Stk_L \approx O(1)$, these particles will be entrained in the fluctuation motion by large-scale eddies of carrier gas and take up the energy from the latter eddies. As a result, the intensity of turbulent fluctuations of the continuous phase may decrease significantly with increasing concentration of particles. A decrease in gas fluctuations will cause some laminarization of turbulent flow, which will result in a less flat profile of averaged velocity of the gas phase of heterogeneous flow.

Nonequilibrium Flow

Heterogeneous flows of this type are most complex from the standpoint of both mathematical and physical simulation. Such flows are most frequently encountered in nature and find practical application: this explains special interest shown in them by researchers.

Maeda et al. [24, 25] studied a developed upward turbulent flow of air in pipes of diameter $D = 38\,\mathrm{mm}$ and $D = 56\,\mathrm{mm}$ in the presence of spherical particles of glass ($d_p = 45\,\mu\mathrm{m}$ and $d_p = 136\,\mu\mathrm{m}$) and copper ($d_p = 93\,\mu\mathrm{m}$). The mass flow rate concentration of the dispersed phase was varied in the range $M_G = 0.1–0.54$. The velocity of carrier air was $U_{xc} = 4.1–5.7\,\mathrm{m\,s}^{-1}$. Experiments revealed that the profile of averaged velocity of particles was flatter as compared to the respective profile for the carrier phase. The flatness of the velocity profile increased with the inertia of particles. It was further revealed that, in the case of values of concentration $M_G \geq 0.3$, the presence of particles resulted in a significant flattening of the profile of averaged velocity of the gas phase. This effect increased further with increasing inertia of the dispersed phase.

Lee and Durst [21] investigated a fully developed upward turbulent flow of air in a pipe of diameter $D = 42\,\mathrm{mm}$ with glass particles ($d_p = 100, 200, 400$, and $800\,\mu\mathrm{m}$). The mass flow rate concentration of the dispersed phase was varied in the range from $M_G = 1.2$ for small particles to $M_G = 2.5$ for large particles. The velocity of carrier air was $U_{xc} = 5.7\,\mathrm{m\,s}^{-1}$. Investigations revealed that the large particles make flatter the profile of averaged velocity of gas by causing its value to decrease in the vicinity of the axis and to increase in the vicinity of the wall.

The most integrated study of nonequilibrium heterogeneous flows in pipes is described in [35, 36, 38, 43]. Tsuji et al. [35, 36] performed measurements for a flow with large particles.

An example of the data obtained by Tsuji et al. [36] on the distributions of averaged velocities of "pure" air and of air in the presence of particles ($d_p = 200\,\mu\mathrm{m}$, $\rho_p = 1,000\,\mathrm{kg\,m}^{-3}$) for an upward turbulent flow in a pipe of

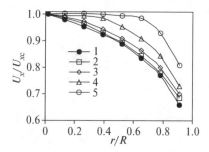

Fig. 4.1. The distribution of averaged velocities of pure air and of air with plastic particles in an upward turbulent flow in a pipe ($U_{xc} \approx 13\,\mathrm{m\,s^{-1}}$, $Re_D \approx 2.3 \times 10^4$): $(1)M = 0$, $(2)M = 0.5$, $(3)M = 1.3$, $(4)M = 1.9$, $(5)M = 3.2$

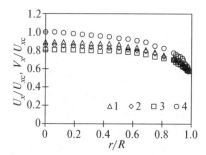

Fig. 4.2. The distribution of averaged velocities of (1–3) solid particles and (4) "pure" air: (1) glass particles (50 µm), (2) alumina particles (50 µm), (3) glass particles (100 µm)

diameter $D = 30.5\,\mathrm{mm}$ is given in Fig. 4.1. By and large, the results given in this figure agree with the inferences made in [21,24,25]. One can see that, in the case of values of concentration $M_G \approx M \geq 1.3$, the presence of particles results in a significant flattening of the profile of averaged velocity of carrier air.

Varaksin et al. [43] investigated an upward turbulent flow of air in a pipe of diameter $D = 64\,\mathrm{mm}$ with nonspherical particles of alumina ($d_p = 50\,\mathrm{\mu m}$, $\rho_p = 3,950\,\mathrm{kg\,m^{-3}}$) and spherical particles of glass ($d_p = 50\,\mathrm{\mu m}$ and $d_p = 100\,\mathrm{\mu m}$). The mass flow rate concentration of particles was varied in the range $\langle M_G \rangle \approx M = 0.12\text{–}0.39$. The averaged velocity of carrier air was $U_{xc} = 6.4\,\mathrm{m\,s^{-1}}$. The measured distributions of averaged velocities of "pure" air and solid particles over the pipe cross-section are given in Fig. 4.2. One can see in this figure that the velocity of solid particles is lower than that of carrier air in almost the entire cross-section of the pipe (except for the wall region), which is natural for the case of upward flow. Note further that no impact of the concentration of particles on their averaged velocity was observed (it was within the experimental error). This was to be expected, because the volume concentration of particles under conditions of this investigation was low $\Phi_G = (3.64\text{–}7.9) \times 10^{-5}$ for alumina particles and $\Phi_G = 5.65 \times 10^{-5} - 1.84 \times 10^{-4}$

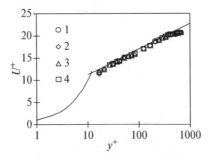

Fig. 4.3. The distribution of averaged velocities of (1–3) the gas phase of heterogeneous flow and (4) "pure" air in universal coordinates: (1) $M = 0.39$, glass particles (50 μm); (2) $M = 0.26$, alumina particles (50 μm); (3) $M = 0.39$ glass particles (100 μm)

for glass particles; this corresponds to the region of dilute heterogeneous flow (see Fig. 1.8) where the interparticle interaction is negligible.

The profiles of averaged velocity of "pure" air and of the air phase of heterogeneous flow are given in universal coordinates in Fig. 4.3. The distributions of velocities of air in the presence of particles correspond to the cases of maximal content of particles in the flow, i.e., $M_G = 0.26$ and $M_G = 0.39$ for flows with particles of alumina and glass, respectively. On analyzing Fig. 4.3, one can conclude that the effect of particles on the distribution of averaged velocity of the carrier phase is negligible. On the one hand, this is attributed to low values of the concentration of the dispersed phase. Another reason is the fact that the particles employed in the experiments exhibited a relatively low inertia; therefore, the profiles of their averaged velocity and of carrier air velocity are similar in shape (see Fig. 4.2). Evidence of this is the fact that the most important characteristic of heterogeneous flow, namely, the averaged Reynolds number of a particle, remained constant for particles of all types in a large region of the pipe ($r/R = 0–0.6$). This characteristic assumed the following values in the above-identified region of flow: $Re_p \approx 2.6$ for alumina particles, and $Re_p \approx 1.7$ and $Re_p \approx 8.5$ for glass particles 50 and 100 μm in diameter, respectively.

Varaksin and Polyakov [38] investigated a downward turbulent flow of air in a pipe of diameter $D = 46$ mm with spherical particles of glass ($d_p = 50$ μm). The mass flow rate concentration of particles was varied in the range $M_G \approx M = 0.05–1.2$. The averaged velocity of carrier air was $U_{xc} = 5.2$ m s^{-1}. The results of measurements of the longitudinal and transverse components of averaged velocities of "pure" air and solid particles over the pipe cross-section are given in Fig. 4.4. One can see in this figure that the longitudinal component of averaged velocity of glass particles is higher than the respective characteristic for carrier air in the entire cross-section of the pipe. This indicates that the stabilization of heterogeneous flow, whose main criterion is the completion of acceleration of particles, has terminated on

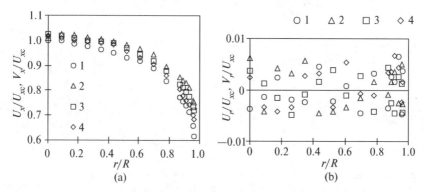

Fig. 4.4. The distributions of (a) longitudinal and (b) transverse components of averaged velocities of (1) "pure" air and (2–4) glass particles in a downward turbulent flow in a pipe ($U_{xc} = 5.2\,\mathrm{m\,s^{-1}}$ $Re_{\mathrm{D}} = 15,300$): (2) $M = 0.05$, (3) $M = 0.35$, (4) $M = 1.2$

reaching the measuring cross-section. Figure 4.4a demonstrates that the profile of the longitudinal component of averaged velocity of particles is flatter as compared to the similar profile for air in the $r/R \leq 0.9$ region. The flatness of the profile of velocity of particles decreases with the increase in their concentration. This is apparently associated with the fact that the increase in the concentration of particles intensifies the interphase exchange of momentum in averaged motion, which leads to the convergence of the profiles of the gas and solid phases.

The profiles of the transverse component of averaged velocities of air and glass particles, given in Fig. 4.4b, clearly demonstrate the fact that the value of this characteristic for both phases of heterogeneous flow is close to zero (the deviations are within the experimental error).

Flow with Large Particles

Heterogeneous flows of this type are characterized by the fact that the relaxation time of particles exceeds significantly the characteristic time of large-scale turbulent eddies, i.e., $Stk_{\mathrm{L}} \to \infty$. Such particles will not react to turbulent fluctuations of the carrier phase velocity, and the distributions of their averaged velocities will be almost uniform over the channel (pipe) cross-section. This observation may be clearly supported by the data of Tsuji et al. [36] on the distributions of averaged velocities of "pure" air and plastic particles ($d_{\mathrm{p}} = 3,000\,\mathrm{\mu m}$, $\rho_{\mathrm{p}} = 1,000\,\mathrm{kg\,m^{-3}}$) over the pipe cross-section, given in Fig. 4.5. This figure may further lead one to infer the presence of significant dynamic slip between the phases in averaged motion. The presence of phase slip leads to intensive exchange of momentum between the gas and particles; this will cause the flattening of the profile of averaged velocity of the carrier phase.

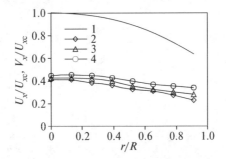

Fig. 4.5. The distributions of averaged velocities of (1) "pure" air and (2–4) plastic particles in an upward turbulent flow in a pipe ($U_{xc} \approx 20\,\mathrm{m\,s}^{-1}$, $Re_\mathrm{D} \approx 3.1 \times 10^4$): (2) $M_\mathrm{G} = 1.2$, (3) $M_\mathrm{G} = 2.2$, (4) $M_\mathrm{G} = 3$

4.2.2 Fluctuation Velocities of Gas and Particles

We will now analyze the presently available experimental data on the distribution of fluctuation (mean- square) velocities of gas in the presence of particles and of the particles proper under conditions of heterogeneous flows of different types.

Equilibrium Flow

As was observed in Sect. 4.2.1, in the case of such a heterogeneous flow, the particles follow completely the turbulent fluctuations of the carrier gas velocity. This is due to the fact that the time of dynamic relaxation of particles is negligible compared to the characteristic time of large-scale turbulent eddies ($Stk_\mathrm{L} \to 0$) and less than (or of the same order with) the time of small-scale eddies, i.e., $Stk_\mathrm{K} \approx O(1)$. As a result, the distribution of fluctuation (mean-square) velocities of extremely low-inertia particles (as well as the distribution of their averaged velocities) will exactly repeat the respective profile for the carrier phase.

QuasiEquilibrium Flow

It was mentioned above that the values of the Stokes number of particles in a flow of this type in large-scale fluctuation motion are of the order of unity, i.e., $Stk_\mathrm{L} \approx O(1)$. Heterogeneous flows of this type are further characterized by the presence of dynamic slip in fluctuation motion. In the case of low values of the Stokes number in large-scale fluctuation motion (say, at $Stk_\mathrm{L} < 0.1$), the particles will have little effect on the intensity of turbulent fluctuations of the carrier phase velocity. In this case, the particles will be entrained in the fluctuation motion by way of spending the energy of high-frequency small-scale eddies whose contribution to the total turbulent energy is small. The effect of the dispersed phase on the energy of turbulent fluctuations of carrier gas will increase with the inertia of particles (their Stokes number Stk_L).

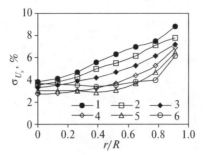

Fig. 4.6. The distributions of the intensity of fluctuations of velocities of "pure" air and air in the presence of plastic particles in an upward turbulent flow in a pipe ($U_{xc} \approx 13\,\mathrm{m\,s^{-1}}$, $Re_D \approx 2.3 \times 10^4$): (1) $M = 0$, (2) $M = 0.5$, (3) $M = 0.9$, (4) $M = 1.3$, (5) $M = 1.9$, (6) $M = 3.2$

Nonequilibrium Flow

The distributions of fluctuation velocities of "pure" air and air in the presence of plastic particles ($d_p = 200\,\mu\mathrm{m}$, $\rho_p = 1,000\,\mathrm{kg\,m^{-3}}$), borrowed from [36], are given in Fig. 4.6. It follows unambiguously from this figure that the presence of relatively low-inertia particles leads to the suppression of the intensity of turbulent fluctuations of carrier air in the entire cross-section of the pipe. As the mass concentration of the dispersed phase increases, this effect increases until $M \approx M_G = 1.3$, and then decreases somewhat at $M_G = 1.9$ and $M_G = 3.2$. At high values of the mass concentration of particles, the profile of fluctuation velocity of air becomes almost uniform in a large region of the pipe ($r/R = 0$–0.8). This is apparently due to the fact that the impact made by the particles on the distribution of averaged velocity of air for the given concentrations of particles becomes significant (see Fig. 4.1). The flatness of the profile of averaged velocity of the carrier phase causes the decrease in the velocity fluctuations and their leveling off over the pipe cross-section. Therefore, the particles affect the intensity of turbulence of gas indirectly by affecting the profile of averaged velocity of the carrier phase.

We will now consider the data of [43] on the profiles of fluctuation velocities of the carrier phase of heterogeneous flow for the case where the presence of particles has no effect on the profile of averaged velocity of gas (see Figs. 4.2 and 4.3). The results of measurements of the longitudinal and transverse components of fluctuation velocity of carrier gas are given in Figs. 4.7 and 4.8, respectively. One can make the following conclusions from Fig. 4.7 (1) all particles employed in the experiment caused a decrease in the intensity of longitudinal fluctuations of carrier air in almost the entire cross-section of the pipe (in the range of $0 \leq r/R \leq 0.9$–0.95); (2) the maximal damping was observed in the vicinity of the pipe axis; and (3) the degree of suppression of longitudinal fluctuations of velocity increases with an increase in the mass concentration of particles and with a decrease in their inertia.

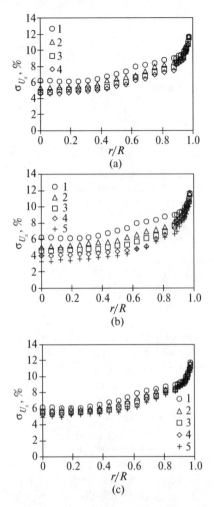

Fig. 4.7. The distributions of the longitudinal component of intensity of fluctuations of velocities of (1) "pure" air and (2–5) the gas phase of heterogeneous flow in the presence of particles of (**a**) alumina (50 μm) ((2) $M = 0.12$, (3) $M = 0.18$, (4) $M = 0.26$); (**b**) glass (50 μm) ((2) 0.12, (3) 0.18, (4) 0.26, (5) 0.39); (**c**) glass (100 μm) ((2) 0.12, (3) 0.18, (4) 0.26, (5) 0.39)

Analysis of the distributions of transverse fluctuations of velocity of carrier air, which are given in Fig. 4.8, reveals the following (1) all particles caused a decrease in the intensity of transverse fluctuations of the gas phase in the entire cross-section of the pipe; (2) the maximal damping of fluctuations was observed in the axial region of the pipe, and the impact made by the presence of particles increases with an increase in their concentration and with a decrease in their inertia; and (3) some tendency is observed for the maximum in the distribution of fluctuations of velocity of the gas phase of heterogeneous

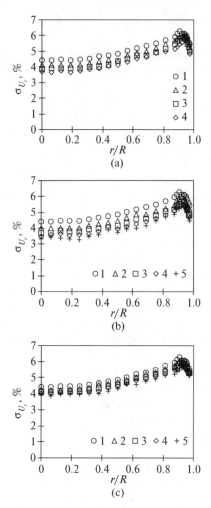

Fig. 4.8. The distributions of the transverse component of intensity of fluctuations of velocities of (1) "pure" air and (2–5) the gas phase of heterogeneous flow in the presence of particles of (**a**) alumina (50 μm) ((2) $M = 0.12$, (3) $M = 0.18$, (4) $M = 0.26$); (**b**) glass (50 μm) ((2) 0.12, (3) 0.18, (4) 0.26, (5) 0.39); (**c**) glass (100 μm) ((2) 0.12, (3) 0.18, (4) 0.26, (5) 0.39)

flow to shift in the direction of the wall compared to the similar distribution for single-phase flow.

The suppression of turbulent fluctuations by particles is largely caused by their involvement in fluctuation motion due to interaction with turbulent eddies of the carrier medium. The higher the degree of involvement of particles in fluctuation motion, the greater their effect on the fluctuation velocity of air. The parameter of dynamic inertia of particles in large-scale fluctuation motion is provided by the respective value of the Stokes number Stk_L.

As the inertia of particles increases, their fluctuation velocity becomes ever lower in the process of interaction with turbulent eddies and, consequently, they take up a smaller amount of turbulent energy from the carrier gas. For example, in the wall region the parameter of particle inertia (Stokes number) increases abruptly compared to its value in the vicinity of the pipe axis because of the decrease in the characteristic lifetime of energy-carrying turbulent eddies (eddies characterized by small size and high values of velocity fluctuations exist in the vicinity of the wall). Therefore, the involvement of particles in fluctuation motion and, consequently, the additional dissipation of turbulence caused by the presence of particles turn out to be less significant than in the vicinity of the pipe axis.

Velocity fluctuations of a steady-state gas flow are defined by the turbulent pattern of flow. As to fluctuations of particle velocities, they may be due to various reasons. Therefore, before analyzing the data on distributions of fluctuation velocities of particles, we will dwell briefly on these reasons which are given schematically in Fig. 4.9. The following fluctuations of velocity of particles moving in a turbulent flow of gas in channels (pipes) may be identified.

First, these are the turbulent fluctuations of velocity of the dispersed phase, which are associated with the involvement of particles in fluctuation motion by turbulent eddies of the carrier phase (they were already mentioned above); secondly, fluctuations of velocity of particles due to their polydispersion, i.e., to the presence of particles of different sizes (and, as a consequence, of different averaged velocities) in the flow; thirdly, fluctuations of velocity of particles because of the variation of their velocities in the process of interaction of particles with one another and with the channel wall; and, fourthly, velocity fluctuations due to migration of particles in a region with a shear of averaged velocity of the dispersed phase.

Figure 4.10 gives profiles of the longitudinal and transverse components of intensity of fluctuations of velocities of "pure" air, air in the presence of particles and particles of glass for different values of their concentration [38]. One can see in the figure that the intensity of longitudinal fluctuations of velocity of particles in the axial region of the pipe is $\sigma_{V_x} = (\overline{v_x'^2})^{1/2}/U_{xc} \approx 8\%$ for a low concentration of particles ($M = 0.05$) and exceeds the respective characteristic for carrier gas (air) $\sigma_{U_x} = (\overline{u_x'^2})^{1/2}/U_{xc} \approx 6\%$. As the concentration of particles increases, the intensity of fluctuations of their velocities in the region identified above decreases to become $\sigma_{V_x} \approx 7\%$ and $\sigma_{V_x} \approx 5\%$ at $M = 0.35$ and $M = 1.2$, respectively. The observed fluctuations of velocity of particles in the vicinity of the pipe axis are caused mainly by their involvement in fluctuation motion by turbulent eddies of the carrier phase and by their polydispersion. The decrease in the intensity of fluctuations of velocity of particles with an increase in their concentration may apparently be explained as follows. The Stokes number in large-scale fluctuation motion is determined as:

$$Stk_L = \frac{\tau_p}{T_L}. \tag{4.4}$$

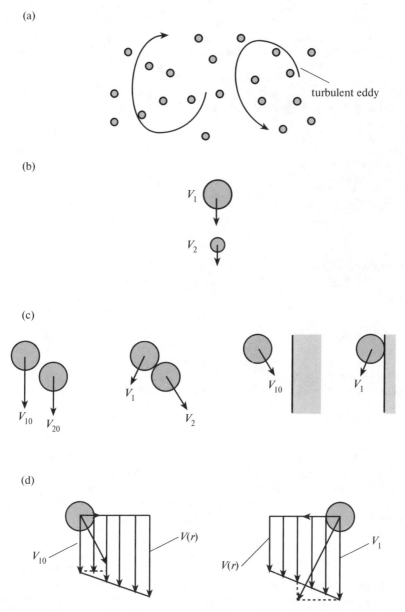

Fig. 4.9. Main reasons for the emergence of fluctuations of velocities of particles in turbulent flows in channels (a) Interaction between particles and turbulent eddies of the carrier phase; (b) Presence of particles of different sizes in the flow; (c) Inter-particle collisions and collisions of particles with the channel wall; (d) Migration of particles in a region with transverse shear of averaged velocity

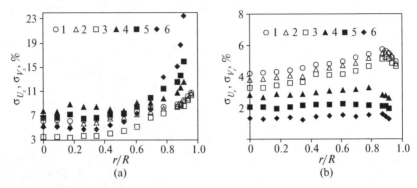

Fig. 4.10. The distributions of (**a**) longitudinal and (**b**) transverse components of intensity of fluctuations of velocities of (1) "pure" air, (2, 3) air in the presence of particles, and (4–6) solid particles in a downward turbulent flow in a pipe ($U_{xc} = 5.2\,\mathrm{ms^{-1}}$, $Re_D = 15,300$) : (2,4) $M = 0.05$, (3, 5) $M = 0.35$, (6) $M = 1.2$

Here, the relaxation time of particles is determined by relation (1.56), and the integral Lagrangian time scale of turbulence which characterizes the lifetime of energy-carrying turbulent eddies may be represented as

$$T_{\mathrm{L}} = \frac{c_\mu^{1/2}k}{\varepsilon} = \frac{l}{c_\mu^{1/4}k^{1/2}}, \tag{4.5}$$

where $c_\mu = 0.09$.

The turbulent energy of the carrier phase is estimated as:

$$2k = \sum_i \overline{u_i'^2} = \overline{u_x'^2} + \overline{u_r'^2} + \overline{u_\varphi'^2} \approx \overline{u_x'^2} + \overline{u_r'^2} + (\overline{u_x'^2} + \overline{u_r'^2})/2. \tag{4.6}$$

For the glass particles employed and for the conditions of the investigation being described, one can use (4.4) in view of (4.5) and (4.6) to derive $Stk_{\mathrm{L}} \approx 1$ for the pipe axis. This implies that the particles are relatively readily involved in large-scale fluctuation motion and, therefore, take up the energy of turbulent eddies of the carrier phase. This effect increases with the concentration of particles. A decrease in the intensity of turbulent fluctuations of the carrier phase leads to a decrease in fluctuations of velocities of solid particles.

In the case of a high concentration of particles, the collisions between them begin to play a decisive part in the formation of the statistical characteristics of the dispersed phase. The interparticle exchange of momentum brings about the leveling off of the velocities of the dispersed phase, which is a factor conducive to a decrease in the intensity of fluctuations of velocities of particles due to their polydispersion.

The longitudinal component of the intensity of fluctuations of particle velocity increases significantly with decreasing distance from the pipe wall (see Fig. 4.10a). At a distance of 2 mm from the wall, which corresponds to the measurement point closest to the wall ($y^+ \approx 20$), the intensity of fluctuations of particle velocity is $\sigma_{V_x} \approx 12\%$ ($M = 0.05$) which is higher than its value for air. The intensity of fluctuations of velocity of particles in the vicinity of the wall increases significantly with the concentration of particles and becomes $\sigma_{V_x} \approx 16\%$ and $\sigma_{V_x} \approx 24\%$ at $M = 0.35$ and $M = 1.2$, respectively. Note that the degree of involvement of particles in large-scale turbulent motion and, consequently, their impact on the intensity of air fluctuations on approaching the wall decrease significantly because of the abrupt increase in the relative inertia of particles (Stokes number Stk_L).

The experimentally observed increase in the fluctuations of particle velocity in this region of the pipe is mainly caused by the high gradients of averaged velocity of the carrier phase (see Fig. 4.4a). The nonuniformity of the air velocity profile further defines the nonuniformity of the profile of averaged velocity of particles which increases with the concentration of particles (see Fig. 4.4a). The particles perform transverse motions as a result of the shear of averaged velocity of air, as well as of the involvement of these particles by turbulent eddies of the carrier phase in the direction normal to the pipe wall. The "free-fall" of particles in a region with different values of averaged velocity of the dispersed phase causes the emergence of high values of fluctuations of particle velocity in the wall region of the pipe.

Excess of fluctuations of velocity of particles over those of the carrier phase was first predicted theoretically by Liljegren [22]. This effect was also revealed in [31,45] in large eddy simulation of particle dynamics in a channel flow and in a flow in a homogeneous shear layer. Excess of fluctuations of particle velocity over those of carrier gas velocity was obtained by Zaichik and Alipchenkov [50] who analyzed the motion of particles in an inhomogeneous turbulent flow using a kinetic equation for the probability density function of velocity of particles. An increase in the intensity of fluctuations of particle velocity on approaching the wall was registered experimentally in [19, 27]. Varaksin et al. [44] also observed excess of longitudinal fluctuations of velocity of glass particles 100 μm in diameter over those of carrier air velocity in almost the entire cross-section of the pipe for a low concentration of the dispersed phase. They revealed a strong dependence of longitudinal fluctuations of particle velocity on the local concentration of the dispersed phase under conditions of significantly nonuniform distribution of the latter phase over the pipe cross-section.

We will consider the data obtained in [38] for the profiles of the transverse component of the intensity of fluctuations of velocities of "pure" air $\sigma_{U_r} = (\overline{u_r'^2})^{1/2}/U_{xc}$ and particles $\sigma_{V_r} = (\overline{v_r'^2})^{1/2}/U_{xc}$. The distributions given in Fig. 4.10b indicate that the intensity of fluctuations of particle velocity in the transverse direction is lower than that for the carrier phase in the entire cross-section of the pipe. The observed fluctuations of particle velocity in the direction being treated are largely caused by the involvement of the dispersed phase in fluctuation motion by turbulent eddies of the carrier phase. The difference in the sizes of particles does not result in the emergence of "additional" fluctuations of velocity of the dispersed phase as in the case of fluctuations in the longitudinal direction described above. This is due to the fact that the transverse component of averaged velocity of particles of different sizes is close to zero (see Fig. 4.4b). An increase in the concentration of particles causes intensification of interphase exchange of momentum in fluctuation motion in the direction being treated. This leads to a decrease in the intensity of fluctuations of carrier air velocity, which, in turn, causes the observed decrease in the fluctuations of particle velocity in the transverse direction. The impact made by the pipe wall consists in that it "interferes" with fluctuations of velocity of the dispersed phase in the direction being treated, which results in the decrease and tendency to zero of the intensity of fluctuations of particle velocity on approaching the wall.

Figure 4.11 gives the distributions of turbulent energy of "pure" air k determined by relation (4.6) and of energy of fluctuations of particle velocity k_p. The approximate estimation of the energy of fluctuations of particle velocity was made using the expression

$$2k_\mathrm{p} = \sum_i \overline{v_i'^2} = \overline{v_x'^2} + \overline{v_r'^2} + \overline{v_\varphi'^2} \approx \overline{v_x'^2} + 2\overline{v_r'^2} \tag{4.7}$$

The assumption of the equality of the tangential (not measured in the experiments) and transverse components of fluctuation velocity of particles $\overline{v_\varphi'^2} \approx \overline{v_r'^2}$, which is made in (4.7), may be argued as follows. The averaged velocity of particles in the tangential direction (as well as in the transverse one) is close to zero. Therefore, the possibility of emergence of "additional" fluctuations of velocity of particles in this direction because of the difference between their averaged velocities due to polydispersion (as in the case of longitudinal fluctuations) is ruled out. In view of the foregoing, possible fluctuations of velocities of particles in the tangential direction are mainly caused by their interaction with turbulent eddies of the carrier phase. The similarity of the intensity of fluctuations of carrier air velocity and their frequency spectrum in the transverse and tangential directions suggests that the respective fluctuations of particle velocity will apparently be similar in magnitude.

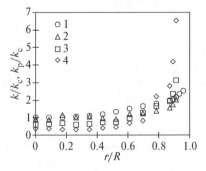

Fig. 4.11. The distributions of (1) turbulent energy of "pure" air and (2–4) energy of fluctuations of particle velocity in a downward turbulent flow: (2) $M = 0.05$, (3) $M = 0.35$, (4) $M = 1.2$

On analyzing the data given in Fig. 4.11, one can conclude that the energy of fluctuations of velocity of particles is lower than the turbulent energy of carrier air in the axial region of the pipe. The energy of fluctuations of particle velocity decreases with increasing concentration of particles in this region of flow. In the wall region of the pipe, the increase in the concentration of the dispersed phase, on the contrary, results in a significant increase in the energy of fluctuations of particle velocity. This characteristic of motion of particles in the vicinity of the wall may exceed significantly the turbulent energy of carrier air.

Flow with Large Particles

We will use the data of Tsuji et al. [36] as an example to illustrate the effect of large particles on the intensity of fluctuations of carrier air velocity. Figure 4.12 gives the results of measurements of distributions of fluctuation velocities of "pure" air and air in the presence of plastic particles ($d_p = 3,000\,\mu m$, $\rho_p = 1,000\,kg\,m^{-3}$ in the pipe cross-section. One can see from these data that the presence of large particles in the flow leads to a significant increase in the intensity of turbulent fluctuations of gas velocity. The observed effect is largely due to the formation of turbulent wakes behind moving particles, which leads to additional generation of turbulence. The effect of generation of fluctuations of gas velocity increases with the concentration of particles and with the distance from the pipe wall.

4.2.3 The Effect of Particles on the Energy Spectrum and Scales of Turbulence of Gas

The data of Tsuji et al. [36] on the impact made by particles on the energy spectrum of turbulent fluctuations of air velocity in the case of nonequilibrium heterogeneous flow are given in Fig. 4.13. One can see that the presence of

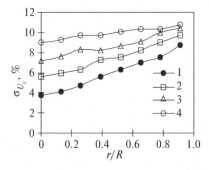

Fig. 4.12. The distributions of the intensity of fluctuations of velocities of (1) "pure" air and (2–4) air in the presence of plastic in an upward turbulent flow in a pipe ($U_{x_c} \approx 13\,m\,s^{-1}$, $Re_D \approx 2.2 \times 10^4$): (2) $M_G = 0.6$, (3) $M_G = 2.3$, (4) $M_G = 3.4$

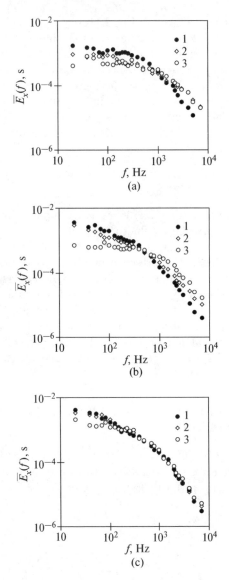

Fig. 4.13. The impact made by the concentration of plastic particles 200 μm in diameter on the energy spectrum of turbulence of carrier air (**a**) $r/R = 0$ ((1) $M_G = 0$, (2) $M_G = 1.3$, (3) $M_G = 3.2$); (**b**) $r/R = 0.521$ ((1) 0, (2) 1.3, (3) 3.2); (**c**) $r/R = 0.912$ ((1) 0, (2) 1.3, (3) 3.2)

plastic particles ($d_p = 200$ μm) in the flow results in a decrease in low-frequency components and increase in high-frequency components of the energy spectrum of turbulence. This effect increases with the concentration of particles and with the distance from the pipe wall.

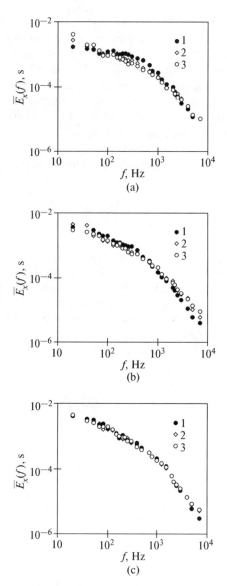

Fig. 4.14. The impact made by the concentration of plastic particles 3,000 μm in diameter on the energy spectrum of turbulence of carrier air (**a**) $r/R = 0$ ((1) M_G = 0, (2) $M_G = 1.1$, (3) $M_G = 3.4$); (**b**) $r/R = 0.521$ ((1) 0, (2) 1.1, (3) 3.4); (**c**) r/R = 0.912 ((1) 0, (2) 1.1, (3) 3.4)

Figure 4.14 gives the results of Tsuji et al. [36] on the effect of the dispersed phase on the energy spectrum of fluctuations for a flow with large particles. These data lead to an unambiguous inference that the presence of large plastic particles ($d_p = 3,000$ μm) in the flow has no effect on the frequency characteristics of turbulence in the entire investigated range of concentrations over the entire cross-section of the pipe. It was mentioned above that experiments revealed a significant increase in the intensity of fluctuations of carrier air velocity in a flow with large particles. We will suggest that the frequency characteristics of turbulence generated in the wake behind particles were close to the respective parameters of "pure" gas; as a result, the presence of particles made no effect on the spectrum of turbulent fluctuations.

We will now turn to treating the results of investigation of the effect of particles on the scales of turbulence of carrier gas. One of the methods of determining the spatial scale is that associated with measuring the distribution of the coefficient of Eulerian time autocorrelation $R_{x,\tau}$ defined by the expression

$$R_{x,\tau} = \frac{\overline{u'_{x,\tau_1} u'_{x,\tau_2}}}{\overline{u'^2_x}}, \qquad (4.8)$$

where u'_{x,τ_1} and u'_{x,τ_2} are the values of the longitudinal component of fluctuation velocity of gas at the instants of time $\tau = \tau_1$ and $\tau = \tau_2$.

Figure 4.15 gives measured distributions of Eulerian time correlation in single-phase and heterogeneous flows [12]. Experiments were performed for a stabilized turbulent flow in a horizontal square channel (55×55 mm^2). The dispersed phase was provided by glass particles ($d_p = 100$–160 μm). The mass concentration of particles was varied in the range $M_G \approx M = 0$–1.0. The dynamic phase slip in averaged motion was 3 ms^{-1}.

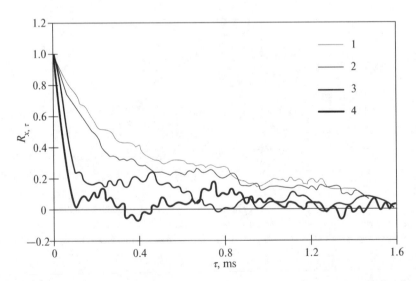

Fig. 4.15. The effect of the concentration of particles on the autocorrelation function in a turbulent flow of air in a horizontal pipe ($U_{xc} = 23$ ms^{-1}, $Re_D = 8 \times 10^4$): (1) $M = 0$, (2) $M = 0.33$, (3) $M = 0.66$, (4) $M = 1$

We will use the distributions of correlation $R_{x,\tau}$ given in Fig. 4.15 to determine the integral time scale of turbulence as follows:

$$T_E = \int_0^\infty R_{x,\tau}(\tau)\mathrm{d}\tau. \tag{4.9}$$

The integral Eulerian time scale of turbulence may be obtained using the relation

$$L_E = U_x \int_0^\infty R_{x,\tau}(\tau)\mathrm{d}\tau = U_x T_E. \tag{4.10}$$

We use relations (4.9) and (4.10) and the data in Fig. 4.15 to determine the scales of turbulence of "pure" air $T_{E_0} = 4.35 \times 10^{-4}\,\mathrm{s}$ and $L_{E_0} = 0.01\,\mathrm{m}$. The presence of particles in the flow results in a significant reduction of the scales of turbulence. For example, we obtained $T_E/T_E = L_E/L_{E_0} = 0.8$ and $T_E/T_{E_0} = L_E/L_{E_0} = 0.15$ for the values of concentration $M = 0.33$ and $M = 1.0$, respectively.

4.2.4 Generalization of Data

Attempts at generalizing the available experimental data for heterogeneous flow of different types in searching for dimensionless parameters which define the effect of solid particles on the flow of carrier gas were made in [9–11, 13, 14]. Gore and Crowe [9] suggested using the ratio of particle diameter to characteristic length scale of flow, d_p/l, as the main dimensionless parameter defining the interaction between particles and turbulence. They demonstrated the existence of the critical value of this parameter, below which the particles suppress the turbulence and above which they generate the turbulence. This critical value is $d_\mathrm{p}/l \approx 0.1$. The length scale of flow (the size of energy-carrying eddies) l was determined from the data of Hutchinson et al. [17]. Hutchinson et al. [17] demonstrated that, for flows in pipes, $l \approx 0.2R$ in the vicinity of the axis and decreases to zero on the wall beginning with $r/R > 0.7$. Gore and Crowe [10] then found that the critical value of d_p/l increases linearly with the distance from the pipe axis and reaches $d_\mathrm{p}/l \approx 0.3$ in the vicinity of the wall. It was mentioned that this parameter provides an answer only to the question of the direction of modification of turbulence (generation or dissipation) rather than to that of the magnitude of this variation.

The effect of another dimensionless parameter, namely, the Reynolds number of particle Re_p determined by the averaged relative velocity between phases, on the interaction between disperse impurity and turbulence of carrier gas was studied by Hetsroni [13, 14]. The data of Gore and Crowe [9] were used to assume that large particles ($Re_\mathrm{p} > 400$) cause eddies behind them, which destabilize the flow and transform the energy of averaged motion to high-frequency components of the energy spectrum of turbulence. Small particles ($Re_\mathrm{p} < 110$) largely suppress the turbulent energy and spend it for their own acceleration (involvement in fluctuation motion). As to medium-sized particles ($110 < Re_p < 400$), they will have a mixed effect on turbulence.

Gore and Crowe [11] continued attempts at finding criteria defining the variation of the intensity of turbulence caused by the presence of the dispersed phase. Nine physical quantities were identified describing the nature of heterogeneous flows, namely:

1. Particle diameter d_p
2. Density of the particle material ρ_p
3. Mass concentration of particles M
4. Density of carrier gas ρ
5. Averaged velocity of gas U_x
6. Dynamic viscosity of gas μ
7. Averaged relative velocity between phases $W = |U_x - V_x|$
8. Characteristic size of energy-carrying eddies l
9. Fluctuation (mean-square) velocity of single-phase flow $(\overline{u_{x_0}'^2})^{1/2}$

The application of the π-theorem led to the following six dimensionless criteria:

$$\frac{(\overline{u_x'^2})^{1/2}}{(\overline{u_{x_0}'^2})^{1/2}} = f\left(\frac{\rho_p}{\rho}, \frac{\rho D U_x}{\mu}, \frac{\rho W d_p}{\mu}, \frac{l}{d_p}, \frac{(\overline{u_{x_0}'^2})^{1/2}}{W}, M\right). \qquad (4.11)$$

Gore and Crowe [11] noted that only four of the foregoing criteria (in which W is absent) may be determined with adequate accuracy without performing experiments. An attempt to combine these criteria into some modified Stokes number $Stk_{Lm} = \rho_p d_p^2 U_x / 18\,\mu l$ and use this number for describing the available experimental data failed to produce a positive result.

The authors of several other studies (in addition to those described above) also tried to find analytically the parameters which define the process of interaction between suspended particles and turbulence of carrier gas [28,29,47–49].

An analytical investigation of additional dissipation of turbulence in a heterogeneous flow containing low-inertia particles was performed by Rogers and Eaton [28] ignoring the volume taken up by the particles. They analyzed the equation for turbulent energy of the carrier phase

$$\frac{Dk}{D\tau} = P - \varepsilon - \frac{1}{\tau_p}\sum_i \left[M(\overline{u_i'u_i'} - \overline{u_i'v_i'})\right.$$
$$\left. + (U_i - V_i)\overline{m'u_i'} + (\overline{m'u_i'u_i'} - \overline{m'u_i'v_i'})\right], \qquad (4.12)$$

where P and ε denote the generation and dissipation of turbulence (they are analogous to the respective terms in the equation for single-phase flow). Rogers and Eaton [28] further performed some simplifications of (4.12) in application to the conditions of boundary layer on a flat plate. First, the fluctuation component of mass concentration has a negligible value for flows with particles of $Stk_L > 1$; as a result, triple correlations of the right-hand part of the equation $\overline{m'u_i'u_i'}$ and $\overline{m'u_i'v_i'}$ are small compared to other terms. Second, the experimental data were used to demonstrate that the averaged relative velocity between particles and carrier gas in the longitudinal direction is of the same order as the fluctuation velocity of gas. The relative velocity in the normal direction has a small value, and the averaged relative velocity in the azimuth direction is almost zero. Therefore, the term $(U_i - V_i)\overline{m'u_i'}$ is of the order of triple correlation and may likewise be ignored.

After simplifications, the equation for turbulent energy (4.12) assumed the form

$$\frac{Dk}{D\tau} = P - \varepsilon - \sum_i \frac{M(\overline{u_i'u_i'} - \overline{u_i'v_i'})}{\tau_{\mathrm{p}}} \qquad (4.13)$$

Rogers and Eaton [27] found that $\overline{u_y'v_y'} \ll \overline{u_y'u_y'}$ and $\overline{u_z'v_z'} \ll \overline{u_z'u_z'}$. It was further observed that $\overline{u_x'v_x'}$ may be a significant part of $\overline{u_x'u_x'}$. As a result, Rogers and Eaton [28] inferred that the additional dissipation of turbulence in a flow with small solid particles will depend on:

1. The averaged value of mass concentration M
2. The time of dynamic relaxation of particle τ_{p}
3. The correlation $\overline{u_x'v_x'}$

Parameters other than those given above, which define the modification of turbulent energy of carrier gas, were found by Sato et al. [29]. They studied upward and downward flows of air with Ni–Zn–ferrite particles ($d_{\mathrm{p}} = 145\,\mu\mathrm{m}$, $\rho_{\mathrm{p}} = 5{,}360\ \mathrm{kgm}^{-3}$) in a $30 \times 80\,\mathrm{mm}^2$ rectangular channel under the effect of magnetic field. This field was generated by two permanent magnets embedded in the wall. The distributions of the longitudinal and normal components of averaged and fluctuation (mean-square) velocities were obtained for both phases of heterogeneous flow in the presence and absence of magnetic field. The measurements revealed that the magnetic field causes an increase in the normal component of averaged and fluctuation velocities of particles. This is the reason for the rise of relative velocity between phases and for the increase in the local concentration of particles in the region where the magnets are located. Analysis of the resultant data, as well as the application of the inferences of Rogers and Eaton [28], resulted in revealing the following four parameters (factors) which define the modification of turbulence for the given experimental conditions:

1. Modified Reynolds number of particle

$$Re_{\mathrm{pm}} - \frac{\sqrt{(V_x - U_x)^2 + (V_y - U_y)^2}\,d_{\mathrm{p}}}{\nu}.$$

2. Ratio of the sums of both components of fluctuation (mean-square) velocity of particles and air

$$\alpha = \frac{(\overline{v_x'^2})^{1/2} + (\overline{v_y'^2})^{1/2}}{(\overline{u_x'^2})^{1/2} + (\overline{u_y'^2})^{1/2}}.$$

3. Effect of local concentration of particles

$$-(U_i - V_i)\overline{m'u_i'} - (\overline{m'u_i'u_i'} - \overline{m'u_i'v_i'}).$$

4. Correlation $\overline{u_i'v_i'}$.

The mathematical model describing the processes of generation and dissipation of turbulence in solid particle-laden flows was suggested by Yarin and Hetsroni [47]. This model is based on the concepts of the pioneering study by Abramovich [1] of the effect of solid particles on the fluctuation velocity of carrier gas. The suggested model rests upon the modified Prandtl mixing length theory and takes into account two main sources of production of turbulence in heterogeneous flows, namely, the gradient of averaged velocity of carrier gas and turbulent wakes behind moving particles. The system of input equations includes (1) the equation of conservation of momentum of individual turbulent eddy and of particles moving in this eddy, (2) the equation of particle motion within a turbulent eddy, and (3) some relations for the flow in the wake behind a particle. The analytical solution of the obtained system of equations resulted in obtaining four dimensionless criteria responsible for the modification of turbulence in heterogeneous flows, namely:

1. The dimensionless diameter of particles $\bar{d}_\mathrm{p} = d_\mathrm{p}/l$
2. The modified Reynolds number of particle $Re_\mathrm{pm} = (\overline{u_{x_0}'^2})^{1/2}d_\mathrm{p}/\nu$
3. The mass concentration of particles M
4. The dimensionless density (density ratio of the phases) $\bar{\rho} = \rho_\mathrm{p}/\rho$

It was further demonstrated by Yarin and Hetsroni [47] that, in the case of flow with relatively small particles ($Re_\mathrm{p} < 400$), the modification of turbulence is defined only by the value of mass concentration of particles M. For a flow with large particles ($Re_\mathrm{p} > 1{,}000$), the variation of turbulent fluctuations will depend on the ratio between mass concentration and dimensionless density $M/\bar{\rho}$, i.e., it will be defined by the volume concentration of disperse impurity Φ.

Comparison of the intensities of turbulence of polydisperse and monodisperse flows was made in [48]. Polydisperse flow was meant a heterogeneous flow in which the dispersed phase is represented by particles of two different sizes. The flow of both types contained relatively small spherical particles, such that the effect of turbulent wakes behind the particles was negligible, i.e., $Re_\mathrm{p} < 110$. The relations derived from the solution of the set of equations analogous to that of [47] resulted in finding:

1. The intensity of turbulence of polydisperse flow is defined by the following parameters – the total mass concentration of the dispersed phase, the mass concentrations of small and large particles, the ratio of the diameters of large and small particles, the density ratio of the phases, the values of the Reynolds number for small and large particles, and the ratio of the mixing length to the diameter of large (small) particles.

2. An increase in the total mass concentration leads to a decrease in turbulent fluctuations of carrier gas velocity.
3. The intensity of turbulence of polydisperse flow may be higher or lower than that of monodisperse one, depending on the ratios of diameters and mass concentrations of large and small particles.
4. In the case of equality of mass concentrations of polydisperse and monodisperse flows, as well as of mass concentrations of large and small particles of polydisperse flow, the intensity of turbulence of polydisperse flow will be higher if the diameter of large particles is larger than the diameter of particles of monodisperse flow and lower if the diameter of large particles is smaller than the diameter of particles of monodisperse flow.
5. If all other parameters are constant, an increase in the mass concentration of small particles leads to an increase in additional dissipation of turbulence, while an increase in the mass concentration of large particles, on the contrary, leads to a decrease in dissipation of turbulence of polydisperse flow.

Yuan and Michaelides [49] developed a simplest theoretical model which takes into account the processes of additional generation and additional dissipation of turbulent energy in heterogeneous flows with large and small particles, respectively. We will not present the final expression for determining the value of turbulent energy of carrier gas in the presence of particles, which was obtained by Yuan and Michaelides [49], because some mathematical inaccuracies were committed in deriving this expression. Note, however, two useful qualitative inferences concerning the suppression and production of turbulent energy in flows with extremely small ($\tau_\mathrm{p} \ll \tau$) and extremely large ($\tau_\mathrm{p} \gg \tau$) particles. Here, the time τ is the least of two times, namely, the lifetime of turbulent eddy and the time of residence of a particle in an eddy. It was found that the dissipation of turbulence in the case of very small particles is proportional to the cube of particle diameter, and the generation of turbulence is proportional to the square of this diameter.

The foregoing description of the results of investigations of particle-laden flows in channels leads one to conclude that the question of finding dimensionless parameters which define the effect of particles on the flow of the carrier phase still remains open.

4.3 Simulation of the Effect of Particles on Turbulent Energy of Gas

It is the objective of this section to analyze separately the laminarizing (dissipative) effect of finely divided impurity [40], the turbulizing effect of large particles due to formation of a wake [52], and the combined effect of both mechanisms on a turbulent flow of gas in a pipe [41]. The physical mechanisms of additional dissipation and additional generation of turbulence in heterogeneous flows are shown schematically in Fig. 4.16.

Spending the energy of eddies
to involve particles
in fluctuation motion

Formation of turbulent
wakes behind
large particles

Additional dissipation
of turbulence

Additional generation
of turbulence

Fig. 4.16. Basic physical mechanisms of modification of turbulent energy in gas flows with low-inertia and large particles

4.3.1 The Dissipation of Turbulent Energy by Small Particles

The additional dissipation of turbulent energy of carrier flow, which is due to the presence of particles, is defined by the relation

$$\varepsilon_{\mathrm{p}} = \frac{2M(k - k_{\mathrm{d}})}{\tau_{\mathrm{p}}}, \tag{4.14}$$

where $k = \frac{1}{2}\sum_i \overline{u_i' u_i'}$ and $k_{\mathrm{d}} = \frac{1}{2}\sum_i \overline{u_i' v_i'}$ denote the turbulent energy of carrier gas and the kinetic energy of interphase interaction, respectively.

Within the locally uniform approximation, which is valid for relatively small particles ($Stk_{\mathrm{L}} = \tau_{\mathrm{p}}/T_{\mathrm{Lp}} < 1$) [51], the kinetic energy of interphase interaction is directly related to the turbulent energy of carrier flow by

$$k_{\mathrm{d}} = f_{\mathrm{u}} k, \quad f_{\mathrm{u}} = \frac{1}{\tau_{\mathrm{p}}} \int_0^\infty \Psi_{\mathrm{u}}(\xi) \exp(-\xi/\tau_{\mathrm{p}}) \mathrm{d}\xi, \tag{4.15}$$

where $\Psi_{\mathrm{u}}(\xi)$ is the time autocorrelation function of fluctuations of gas velocity along trajectories of particles.

We preassign $\Psi_{\mathrm{u}}(\xi)$ in the form of two-scale parabolic–exponential function [6]

$$\Psi_{\mathrm{u}} = \begin{cases} 1 - \xi^2/\tau_{\mathrm{Tp}}^2, & \xi \leq \xi_0, \\ \dfrac{2\xi_0 T_{\mathrm{Lp}}}{\tau_{\mathrm{Tp}}^2} \exp(\frac{\xi_0 - \xi}{T_{\mathrm{Lp}}}), & \xi > \xi_0, \end{cases} \tag{4.16}$$

which satisfies the conditions $\Psi_{\mathrm{u}}(\xi_0 - 0) = \Psi_{\mathrm{u}}(\xi_0 + 0)$ and $\Psi_{\mathrm{u}}'(\xi_0 - 0) = \Psi_{\mathrm{u}}'(\xi_0 + 0)$ at $\xi_0 = \left(\sqrt{1 + Z_{\mathrm{p}}^2} - 1\right) T_{\mathrm{Lp}}$, where $Z_{\mathrm{p}} = \tau_{\mathrm{Tp}}/T_{\mathrm{Lp}}$.

Here, T_{Lp} is the integral time macroscale of fluctuations of gas velocity which is calculated along trajectories of particles and characterizes the time of their interaction with energy-intensive turbulent eddies of carrier flow, and τ_{Tp} is the differential time microscale which characterizes the time of interaction of particles with small-scale turbulent motion. In the case of small particles, when the effects of "inertia" and "crossing of trajectories" may be ignored [46], T_{Lp} coincides with the integral Lagrangian scale of turbulence at infinitely high values of the Reynolds number T_{L}, and τ_{Tp} becomes equal to the Taylor time microscale for fluctuations of gas velocity τ_{T}. The integral time scale of turbulence is defined by the relation

$$T_{\mathrm{L}} = \frac{C_{\mu}^{1/2} k}{\varepsilon}, \tag{4.17}$$

where ε is the rate of dissipation of turbulent energy, and the constant $C_{\mu} = 0.09$.

The quantity $Z = \tau_{\mathrm{T}}/T_{\mathrm{L}}$ which characterizes the ratio of micro- and macroscales is a function of the Reynolds number, and $Z \to 0$ at $Re \to \infty$.

The coefficient of involvement of particles in turbulent motion f_{u} in (4.15) in view of expression (4.16) for the autocorrelation function takes the form

$$f_{\mathrm{u}} = 1 + \frac{2 Stk_{\mathrm{L}}^2}{Z_{\mathrm{p}}^2} \left[\frac{Stk_{\mathrm{L}} + \sqrt{1 + Z_{\mathrm{p}}^2}}{1 + Stk_{\mathrm{L}}} \exp\left(-\frac{\sqrt{1 + Z_{\mathrm{p}}^2} - 1}{Stk_{\mathrm{L}}} \right) - 1 \right]. \tag{4.18}$$

In accordance with (4.15) and (4.18), the additional dissipation of turbulent energy (4.14) is represented as

$$\varepsilon_{\mathrm{p}} = \frac{2 M k \varDelta_{\mathrm{p}}}{T_{\mathrm{Lp}}}, \quad \varDelta_{\mathrm{p}} = \frac{1 - f_{\mathrm{u}}}{Stk_{\mathrm{L}}}, \tag{4.19}$$

where $\varDelta_{\mathrm{p}}$ is the additional dissipation factor,

$$\varDelta_{\mathrm{p}} = \frac{2 Stk_{\mathrm{L}}}{Z_{\mathrm{p}}^2} \left[-\frac{Stk_{\mathrm{L}} + \sqrt{1 + Z_{\mathrm{p}}^2}}{1 + Stk_{\mathrm{L}}} \exp\left(-\frac{\sqrt{1 + Z_{\mathrm{p}}^2} - 1}{Stk_{\mathrm{L}}} \right) + 1 \right]. \tag{4.20}$$

At $Z_{\mathrm{p}} \to 0$ ($Re \to \infty$), (4.18) yields the known expression for the coefficient of involvement of particles in large-scale turbulent motion

$$f_{\mathrm{u}} = \varDelta_{\mathrm{p}} = (1 + Stk_{\mathrm{L}})^{-1}, \tag{4.21}$$

which corresponds to preassigning the autocorrelation function in the form of the exponential dependence

$$\Psi_{\mathrm{u}} = \exp\left(-\frac{\xi}{T_{\mathrm{Lp}}} \right) \tag{4.22}$$

The exponential function (4.22) describes well the behavior of the autocorrelation function at high values of the Reynolds number except for the neighborhood of $\xi = 0$, because it does not meet the condition of symmetry $\Psi_u'(0) = 0$. As a result of this defect in the behavior of Ψ_u at $\xi \to 0$, the additional dissipation (4.19) in determining the dissipation factor according to (4.21) does not go to zero for inertialess particles ($\tau_p \to 0$) but tends to a finite limit. Therefore, formula (4.21) for the additional dissipation factor Δ_p (as well as for the coefficient of involvement f_u) may be used only for particles whose relaxation time exceeds the time microscale of turbulence.

Figure 4.17 gives the additional dissipation factor Δ_p as a function of the Stokes number Stk_L. One can see that the function $\Delta_p = \Delta_p(Stk_L)$ is characterized by the presence of a maximum whose position tends to the point $Stk_L = 0$ as Z_p decreases. At $Z_p \ll 1$, the value of this maximum is close to unity and corresponds to the point $Stk_L = Z_p^2/2$. Therefore, formula (4.21) for Δ_p turns out to be valid for particles whose parameter of inertia satisfies the condition $Stk_L \gg Z_p^2/2$.

The effect of small particles on the turbulence of carrier flow may be characterized by the ratio of additional dissipation of turbulent energy ε_p to viscous dissipation ε. As applied to a pipe flow, the value of ε is determined from the relation

$$\varepsilon = \frac{C_\mu^{3/4} k^{3/2}}{l}, \tag{4.23}$$

where l is the Prandtl–Nikuradse mixing length,

$$l = 0.4y(1 - 1.1\bar{y} + 0.6\bar{y}^2 - 0.15\bar{y}^3), \quad \bar{y} = y/R. \tag{4.24}$$

It is assumed that the particles are rather small, so that the effects of "inertia" and "crossing of trajectories" may be ignored for the time of their interaction with energy-intensive turbulent eddies; therefore, T_{Lp} is taken to be equal to T_L. We will further omit from analysis the case of very small particles which do not obey the condition $Stk_L \gg Z^2/2$. Then, in view of (4.17), (4.19), (4.21), and (4.23),

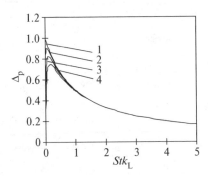

Fig. 4.17. The additional dissipation factor as a function of the parameter of inertia of particles: (1) $Z_p = 0$, (2) 0.1, (3) 0.2, (4) 0.3

$$\frac{\varepsilon_{\mathrm{p}}}{\varepsilon} = \frac{2M \Delta_{\mathrm{p}}}{C_{\mu}^{1/2}} = \frac{2M}{C_{\mu}^{1/2} \left(1 + \frac{C_{\mu}^{1/4} \overline{\tau}_{\mathrm{p}} \overline{k}^{1/2}}{\overline{l}} \right)}, \tag{4.25}$$

where $\overline{\tau}_{\mathrm{p}} = \tau_{\mathrm{p}} u_{*0}/R$, $\overline{k} = k/u_{*0}^2$, $\overline{l} = l/R$ and u_{*0} is the dynamic velocity (friction velocity) in the absence of particles in the flow.

According to (4.25), the effect of relatively small particles on turbulence is defined primarily by the mass concentration M and dimensionless relaxation time $\overline{\tau}_{\mathrm{p}}$. As the inertia of particles decreases (to a certain extent), the laminarizing effect of the dispersed phase on the flow increases. The parameter l in (4.25) allows for the increase in the effect of the dispersed phase with increasing distance from the wall; this is attributed to the decrease in the relative inertia of particles with increasing time scale of turbulence.

4.3.2 The Generation of Turbulent Energy by Large Particles

Semiempirical reasoning is used in [5, 18, 49] to introduce into the equation of balance of turbulent energy of continuous carrier phase additional terms due to the generation of turbulent fluctuations of velocity at high values of the Reynolds number of the flow past particles. Yarin and Hetsroni [47] used directly the self-similar solution for the far axisymmetric turbulent wake [2] in order to estimate the turbulization of flow by large particles. Naturally, this approach is valid only in the case of a very low volume concentration of the dispersed phase, in the absence of interference of wakes behind individual particles. Rather than employing the solution for self-similar turbulent wake for direct calculation of the turbulent characteristics of carrier flow, I use this solution for determining the additional generation of turbulence in the balance equation for fluctuation energy. This interpretation of self-similar solution (i.e., using this solution locally rather than integrally) makes the suggested model valid for various two-phase turbulent flows and gives reason to hope for its validity under conditions of both low and moderate volume concentrations of particles.

The distribution of averaged velocity in a self-similar axisymmetric turbulent wake behind a body (particle) subjected to flow is described as follows [2]:

$$\frac{U_{\delta} - U_{x}}{U_{\delta} - U_{0}} = f(\eta), \tag{4.26}$$

where

$$f(\eta) = (1 - \eta^{3/2})^2, \quad \eta = y/\delta, \quad \delta = 3(\beta^2 C_{\mathrm{D}} d_{\mathrm{p}}^2/16A)^{1/3} x^{1/3},$$

$$A = \int_0^1 f(\eta)\eta \mathrm{d}\eta = \frac{9}{70}, \quad U_{\delta} - U_{0} = \frac{U_{\delta}}{9}(C_{\mathrm{D}} d_{\mathrm{p}}^2/16A\beta^4)^{1/3} x^{-2/3}.$$

Here, x and y are coordinates in the longitudinal and transverse directions to the main flow, U_0 is the velocity on the axis ($y = 0$), U_{δ} is the velocity of unperturbed flow outside of the wake limits, δ is the wake half-width, d_p is the particle diameter, and C_D is the particle drag coefficient. The constant β is defined as the ratio of the mixing length l to the wake half-width ($l = \beta\delta$) and is taken to be equal to 0.2.

In order to calculate the turbulent characteristics of flow in a wake, we will use the balance equation for turbulent energy in a diffusionless approximation, i.e., assuming generation to be equal to dissipation

$$-\overline{u'_x u'_y}\frac{\partial U_x}{\partial y} = \varepsilon. \tag{4.27}$$

The Reynolds shear stress is determined using the Kolmogorov–Prandtl relation,

$$-\overline{u'_x u'_y} = C_\mu \frac{k^2}{\varepsilon}\frac{\partial U_x}{\partial y}. \tag{4.28}$$

Equations (4.27) and (4.28) in view of (4.23) give the following for the rate of dissipation of turbulent energy:

$$\overline{u'_x u'_y} = -\left(\beta\delta\frac{\partial U_x}{\partial y}\right)^2, \quad k = \frac{1}{C_\mu^{1/2}}\left(\beta\delta\frac{\partial U_x}{\partial y}\right)^2, \quad \varepsilon = (\beta\delta)^2\left(\frac{\partial U_x}{\partial y}\right)^3. \tag{4.29}$$

In accordance with (4.29), the distribution of generation of turbulent energy over the cross-section of the wake behind a particle is defined by the expression

$$P = -\overline{u'_x u'_y}\frac{\partial U_x}{\partial y} = (\beta\delta)^2\left(\frac{\partial U_x}{\partial y}\right)^3$$

or, in view of (4.26),

$$P = -\frac{\beta^2(U_\delta - U_0)^3}{\delta}f'^3(\eta) = \left(\frac{C_D d_p^2}{16A\beta^4}\right)^{2/3}\times\frac{U_\delta^3}{81x^{7/3}}\eta^{3/2}(1-\eta^{3/2})^3 \tag{4.30}$$

We will now calculate the additional generation of turbulent energy in the volume of a cell per particle. We assume the distribution given by (4.30) to be valid up to the particle surface and derive

$$P_p = \frac{1}{\Omega}\int P d\omega = \frac{2\pi}{\Omega}\int_{d_p/2}^{d_c/2}\int_0^\delta P y\, dy\, dx$$
$$= \frac{B}{27\times 2^{11/3}}\left(\frac{C_D}{A\beta}\right)^{4/3}\Phi(1-\Phi^{2/9})\frac{U_\delta^3}{d_p}, \tag{4.31}$$

where $B = -\int_0^1 f'^3(\eta)\eta\, d\eta = 0.6$.

Here $\Omega = N_p^{-1} = \pi d_c^3/6$ is the cell volume, N_p is the number of particles per unit volume, d_c is the conditional diameter of cell, and $\Phi = \pi d_p^3 N_p/6$.

Strictly speaking, formula (4.31) is valid for low volume concentrations of particles ($\Phi \ll 1$). Therefore, the additional generation of turbulence in the wake behind particles will be finally represented as

$$P_{\mathrm{p}} = a \left(\frac{C_{\mathrm{D}}}{\beta} \right)^{4/3} \Phi \frac{W^3}{d_{\mathrm{p}}}, \quad a = 0.027, \qquad (4.32)$$

where U_δ is replaced by $W = |U_x - V_x|$ (which corresponds to the modulus of relative averaged velocity of the continuous and dispersed phases).

Note that (4.32) turns out to be rather close in form to the expression for additional generation of turbulent fluctuations at high values of the Reynolds number of flow past particles, which was obtained semiempirically by Derevich [5].

The ratio of P_{p} to viscous dissipation ε may be used as the quantity characterizing the effect of additional generation in the wake behind particles on the turbulence of carrier flow,

$$\frac{P_{\mathrm{p}}}{\varepsilon} = \frac{a}{C_\mu^{3/4}} \left(\frac{C_{\mathrm{D}}}{\beta} \right)^{4/3} \frac{\Phi \overline{W}^3 \overline{l}}{\overline{d}_{\mathrm{p}} \overline{k}^{3/2}}, \qquad (4.33)$$

where $\overline{W} = W/u_{*0}$ and $\overline{d}_{\mathrm{p}} = d_{\mathrm{p}}/R$

One can see from (4.33) that the effect of large particles on turbulence is largely defined by the volume concentration Φ (rather than by the mass concentration M, as in the case of finely divided impurity), the dimensionless velocity of interphase slip $\overline{W}$, and the dimensionless diameter $\overline{d}_{\mathrm{p}}$. The turbulizing effect of the dispersed phase on the flow increases with the particle size (because the value of $C_{\mathrm{D}}^{4/3} \overline{W}^3 / \overline{d}_{\mathrm{p}}$ increases). The parameter $\overline{l}$ in (4.33) indicates that the turbulizing effect of large particles, as well as the laminarizing effect of small particles in accordance with (4.25), increases with the distance from the wall.

4.3.3 The Effect of Particles on Turbulent Energy of Gas

We will consider the effect of the dispersed phase on the turbulence intensity of a steady-state hydrodynamically developed flow in a vertical round pipe. The balance equation for turbulent energy is used to analyze the effect of particles on the level of turbulence,

$$\frac{\mathrm{D}k}{\mathrm{D}\tau} = D + P - \varepsilon + P_{\mathrm{p}} - \varepsilon_{\mathrm{p}}. \qquad (4.34)$$

The term on the left-hand side of (4.34) denotes the time variation and convective transfer of turbulent energy. The terms on the right-hand side, respectively, describe the diffusion transfer due to fluctuations of velocity and pressure, the generation of turbulence from averaged motion, the dissipation of turbulent energy, and the additional generation and dissipation due to the presence of particles in the flow. In the treated case of steady-state hydrodynamically developed flow, the left-hand part of (4.34) vanishes. Further, with a view to deriving a simple analytical relation for the effect of particles on turbulence, we will perform analysis in the so-called diffusionless (algebraic) approximation, i.e., disregarding the contribution by the diffusion term in (4.34).

The generation of turbulent energy in (4.34) due to the shear of averaged velocity is defined by the relation

$$P = -\overline{u'_x u'_y}\frac{\partial U_x}{\partial y} = C_\mu \frac{k^2}{\varepsilon}\left(\frac{\partial U_x}{\partial y}\right)^2, \tag{4.35}$$

where x is the longitudinal coordinate along the pipe axis, and y is the transverse coordinate directed from the wall toward the pipe axis.

The contribution of additional dissipation ε_p to the balance of turbulent energy (4.34) is significant only for relatively small particles ($Stk_L < 1$). Therefore, the time of interaction between particles and turbulent eddies T_{Lp} may be taken to be equal to the time scale of turbulence T_L, because the error in determining T_{Lp} plays no significant part due to the small contribution of ε_p to (4.19) for large particles. Then, (4.34) in view of (4.17), (4.19), (4.23), (4.32), and (4.35) yields the following expression for the turbulent energy of carrier flow:

$$k = \frac{\dfrac{l^2}{C_\mu^{1/2}}\left(\dfrac{\partial U_x}{\partial y}\right)^2 + \dfrac{a}{C_\mu^{3/4}}\left(\dfrac{C_D}{\beta}\right)^{4/3}\dfrac{\Phi W^3 l}{k^{1/2}d_p}}{1 + \dfrac{2M\Delta_p}{C_\mu^{1/2}}}. \tag{4.36}$$

We will assume that the impact of the dispersed phase on the profile of averaged velocity of gas, as well as that on the distribution of mixing length, may be ignored in analyzing the effect of the dispersed phase on the intensity of turbulent energy (in a first approximation). In addition, we will restrict ourselves to treating the particles which satisfy the condition $Stk_L \gg Z^2/2$. In this case, expression (4.36) in view of (4.21) may be represented as

$$\frac{k}{k_0} = \frac{1 + \dfrac{a}{C_\mu^{3/4}}\left(\dfrac{C_D}{\beta}\right)^{4/3}\dfrac{\Phi \overline{W}^3 \overline{l}}{\overline{k_0}^{3/2}\overline{d_p}}\left(\dfrac{k_0}{k}\right)^{1/2}}{1 + 2M/(C_\mu^{1/2}[1 + C_\mu^{1/4}(\overline{\tau}_p \overline{k_0}^{1/2}/\overline{l})(k/k_0)^{1/2}])}, \tag{4.37}$$

where $\overline{k}_0 = k_0/u_{*0}^2$ and k_0 is the turbulent energy in the absence of particles from the flow.

In the case of small particles, when the additional generation of turbulence in the wake behind particles is of no importance, (4.37) yields

$$\frac{k}{k_0} = \frac{1}{1 + 2X/C_\mu^{1/2}}, \tag{4.38}$$

where $X = \dfrac{M}{1 + C_\mu^{1/4}(\overline{\tau}_p \overline{k_0}^{1/2}/\overline{l})(k/k_0)^{1/2}}.$

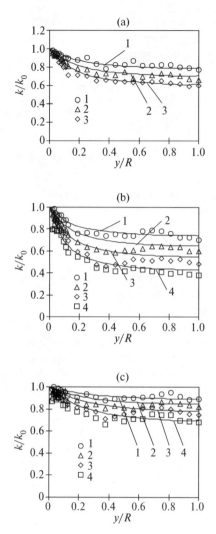

Fig. 4.18. The effect of low-inertia particles on the distribution of turbulent energy of gas over the cross-section of a pipe (**a**) alumina (50 μm), (**b**) glass (50 μm), (**c**) glass (100 μm); lines 1–4 – formula (4.38): (1) $M = 0.12$, (2) 0.18, (3) 0.26, (4) 0.39

In Fig. 4.18, relation (4.38) is compared with the experimental data of [43] for small particles on the distribution of the turbulent energy of gas related to the respective value obtained in the absence of particles from the flow. The transversal component of velocity fluctuations (not measured in the experiments) was calculated in terms of the longitudinal and transverse components using the relation $\overline{u'_\varphi{}^2} = (\overline{u'_x{}^2} + \overline{u'_r{}^2})/2$. One can see in Fig. 4.18

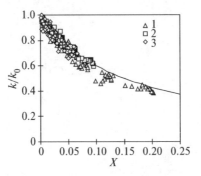

Fig. 4.19. The effect of low-inertia particles on the intensity of turbulence: (1) glass (50 μm), (2) alumina (50 μm), (3) glass (100 μm). The *line* indicates calculation by formula (4.38)

that, indeed, the presence of relatively small particles in the flow causes the suppression of turbulence. In so doing, the laminarizing effect of the dispersed phase in accordance with (4.38) increases with the mass concentration of particles and with the distance from the wall. As the inertia of particles increases (in the range being treated), their impact on turbulence decreases.

Figure 4.19 gives all experimental data from Fig. 4.18, generalized in the coordinates k/k_0 with respect to X. One can infer from Fig. 4.19 that the experimental data are grouped together fairly densely in the form of dependence of k/k_0 on X and are described adequately by formula (4.38). Note that formula (4.38) does not satisfy the obvious passage to the limit for the case of inertialess particles, namely, $k \to k_0$ at $\tau_p \to 0$. Therefore, as was already mentioned, formula (4.38) is valid for analysis of the effect made on turbulence by particles whose relaxation time exceeds the time microscale. Relation (4.20) may be used to provide for a correct passage to the limit at $\tau_p \to 0$ for the purpose of calculating the additional dissipation factor.

In the case of large particles, where the additional dissipation of turbulence is insignificant, formula (4.37) yields

$$\frac{k}{k_0} = 1 + \left(\frac{k_0}{k}\right)^{1/2} bY,$$

$$b = \frac{a}{C_\mu^{3/4}\beta^{4/3}}, \tag{4.39}$$

$$Y = \frac{C_D^{4/3}\varPhi\overline{W}^3\overline{l}}{\overline{d}_p\overline{k}_0^{3/2}}.$$

In Fig. 4.20, relation (4.39) is compared with the experimental data of [36] for large particles on the distribution of turbulence intensity related to the

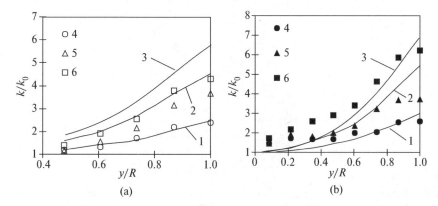

Fig. 4.20. Comparison of (1–3) results of calculation by formula (4.39) and (4–6) experimental results of investigation of the effect of large plastic particles on the distribution of turbulent energy of gas over the pipe cross-section (**a**) 1,500 μm (($(1, 4)\Phi = 0.00071$, (2, 5) 0.0023, (3, 6) 0.0035); (**b**) 3,000 μm (($(1, 4)$ $\Phi = 0.00071$, (2, 5) 0.0027, (3, 6) 0.004)

respective value obtained in a single-phase flow over the pipe cross-section under conditions of upward flow. Because Tsuji et al. [36] measured only the axial component of velocity fluctuations, their relative value was taken to be equal to relative turbulent energy. One can see in Fig. 4.20 that, in accordance with (4.39), the turbulizing effect of the large-particle dispersed phase increases with the volume concentration and particle size, as well as with the distance from the wall. Therefore, as in the case of finely divided impurity, the wall region turns out to be more conservative (less sensitive) compared to the flow core as regards the effect of the dispersed phase on the turbulent structure of carrier flow.

Figure 4.21 gives the experimental data of Fig. 4.20 generalized in the co-ordinates k/k_0 with respect to Y. One can see that the experimental data are generalized fairly well in the form of dependence of k/k_0 on Y and are described adequately by formula (4.39).

We will perform a qualitative analysis of the effect of particles on the turbulence of carrier flow under conditions of flow in a vertical pipe. The averaged velocity of interphase slip will be estimated as

$$W = \tau_p g, \tag{4.40}$$

where g is the acceleration of gravity, and the time of dynamic relaxation of particles is defined by the known relation (1.56),

$$\tau_p = \tau_{p0}/C(Re_p), \quad \tau_{p0} = \rho_p d_p^2/18\rho\nu, \quad Re_p = Wd_p/\nu. \tag{4.41}$$

The particle drag coefficient is

$$C_D = 24C(Re_p)/Re_p. \tag{4.42}$$

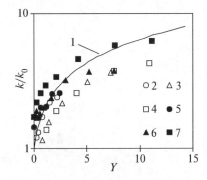

Fig. 4.21. The effect of large plastic particles on the turbulent energy of gas: (1) formula (4.39), (2–4) 1,500 μm, (5–7) 3,000 μm

Relations (4.40) and (4.41) will be written in dimensionless form as

$$\overline{W} = \overline{\tau}_{\mathrm{p}}G, \quad \overline{\tau}_{\mathrm{p}} = \frac{\rho_{\mathrm{p}}}{\rho}\frac{\overline{d}_{\mathrm{p}}^2 R_+}{18C(Re_{\mathrm{p}})}, \qquad (4.43)$$

where

$$G = gR/u_{*0}^2, \text{ and } R_+ = Ru_{*0}/\nu.$$

We will restrict ourselves to analyzing the effect of the dispersed phase on the turbulent energy of flow at the pipe center (on the axis), where we can assume

$$\overline{l} = 0.14, \quad \overline{k}_0 = 1.$$

Then, it follows from (4.37) in view of (4.42) and (4.43) that the effect of the dispersed phase on the turbulent energy of carrier flow is represented in the form of the following functional dependence:

$$\frac{k}{k_0} = f(\rho_{\mathrm{p}}/\rho, \Phi, \overline{d}_{\mathrm{p}}, R_+, G). \qquad (4.44)$$

The importance of individual parameters in (4.44) varies with increasing particle size. For example, the importance of the gravity parameter G defining the averaged interphase slip is significant only for large particles. In the case of small particles, the effect of individual parameters on turbulence in accordance with (4.38) shows up in terms of the complexes $M = (\rho_{\mathrm{p}}/\rho)\Phi$ and $\overline{\tau}_{\mathrm{p}} \sim (\rho_{\mathrm{p}}/\rho)\overline{d}_{\mathrm{p}}^2 R_+$. In the case of large particles, in accordance with (4.39), $\Phi, \overline{d}_{\mathrm{p}}$, and $\overline{W} \sim [(\rho_{\mathrm{p}}/\rho)\overline{d}_{\mathrm{p}}G]^{1/2}$ become the determining parameters.

5

Particle-Laden Flows Past Bodies

5.1 Preliminary Remarks

The chapter deals with solid particle-laden flows past bodies. This problem emerged as a result of studies of motion of various flying vehicles in a dust-laden atmosphere, as well as of motion of two-phase heat-transfer agents in flow trains of power plants. The presence of solid particles may cause a significant (sometimes, many times over) increase in heat fluxes, as well as erosion wear of the surface subjected to flow. These phenomena are due to the combined effect of a number of reasons, which include the variation of the structure of flow incident on a body and of the characteristics of the boundary layer developing on the body subjected to flow, as well as particle/surface collisions, variation of the surface roughness, and so on. The intensity of the processes which accompany heterogeneous flows past bodies depends on the inertia and concentration of particles. Note that the inertia of particles depends directly on the geometry and parameters of flow and may vary in a very wide range for the same particles. The presence of different characteristic times (lengths) of carrier flow (in the vicinity of the critical point of the body subjected to flow and along its surface, turbulent scales proper, and so on) complicates seriously the study of such flows and generalization of data. As to the concentration of particles, its value may be many times the "initial" value in an unperturbed flow due to abrupt deceleration of flow on approaching the body, particle/ wall interaction, and interparticle collisions. When particles move along the surface in the boundary layer characterized by significant gradients of velocity and temperature (in the case of nonisothermal flow), their distribution is often complex, and the value of concentration is likewise higher than that in the flow incident on the body.

Studies into heterogeneous flows past bodies are mainly aimed at (1) investigating the motion of particles and determining their trajectories, (2) determining the effect of particles on the flow of gas, and (3) investigating the processes of interaction of particles with the surface subjected to flow, including the erosion of material. Primary consideration will be given to the first two problems.

Described and analyzed below are the results of mathematical and physical simulation of particle-laden flows past bodies. Section. 5.2 is devoted to treatment of the characteristics of dust-laden flows in the region of the critical (frontal) point of a body. The characteristic features and parameters of heterogeneous flows along the surface subjected to flow will be treated in Sect. 5.3. The results of investigations of the drag of bodies subjected to heterogeneous flows are described in Sect. 5.4.

5.2 A Flow with Particles in the Region of the Critical Point of a Body

In order to analyze the physical processes occurring in the region of the frontal point of a body subjected to heterogeneous flow, one must know the distributions of velocities, temperatures (in the case of nonisothermal flow), and concentration of particles. A very appropriate (in my opinion) classification of possible modes of heterogeneous flows past bodies was suggested by Tsirkunov [25]. This classification is given in Fig. 5.1 with the Stokes number ranges in accordance with the classification of particle-laden gas flows suggested in Sect. 1.5.

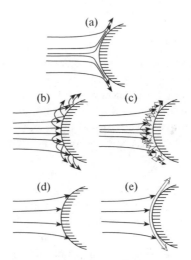

Fig. 5.1. The classification of modes of heterogeneous flows past bodies: (**a**) equilibrium flow, quasiequilibrium flow, $Stk_f \to 0$; (**b**) weakly dust-laden nonequilibrium flow, flow with large particles, $Stk_f \approx O(1)$; (**c**) highly dust-laden nonequilibrium flow, flow with large particles, $Stk_f \approx O(1)$; (**d**) flow past a body in the case of an absorbing wall; (**e**) flow past a body with the formation of a film

Sedimentation factor The sedimentation (trapping) factor of particles is an important integral characteristic of heterogeneous flows past bodies. This factor is the ratio of the number of particles which collided with the body to the number of particles which could collide with the body if their lines of flow were straight lines.

In the case of axisymmetric flow (for example, flow past a sphere), where the size of particle is negligible compared to the size of body, particles are uniformly distributed in the incident flow, and their trajectories are symmetrical, the sedimentation factor may be determined as

$$\eta = \overline{y}_{\text{cr}}^2, \tag{5.1}$$

where $\overline{y}_{\text{cr}} = y_{\text{cr}}/R$ is the dimensionless distance from the symmetry axis of flow, at which the particles (in a flow unperturbed by the presence of a body) only touch the body while flowing past it. The particles, whose coordinates in the incident flow are $\overline{y} > \overline{y}_{\text{cr}}$, do not collide with the body.

In the case of plane flow (for example, transverse flow past an infinite cylinder or plate), expression (5.1) for the sedimentation factor is simplified and takes the form

$$\eta = \overline{y}_{\text{cr}}, \tag{5.2}$$

We will now analyze the results of theoretical and experimental investigations of motion of particles and determination of their effect on the flow of carrier gas in the vicinity of the critical points of bodies subjected to flow.

5.2.1 Theoretical Investigations

One of the early investigations of particle trajectories under conditions of potential heterogeneous flow past a sphere was performed by Michael and Norey [11]. Because the velocities of gas and particles away from the body surface were taken to be equal to each other, the flow away from the body, treated by Michael and Norey [11], may be classified as quasiequilibrium (see Table 1.1). As a result of the difference between the velocities of the phases due to particle inertia in the vicinity of the frontal point, this flow became nonequilibrium. The calculations were performed for the case where the resistance of particles obeys the Stokes law. As to the concentration of particles, a weakly dust-laden flow without inverse effect of particles on the carrier gas was treated. The interaction between the particles and the sphere was not treated, because it was assumed that the particles were absorbed by the surface of the sphere.

The velocity field of gas was defined by the following relations:

$$\overline{U}_x = 1 + \frac{\overline{y}^2 - 2\overline{x}^2}{2(\overline{x}^2 + \overline{y}^2)^{5/2}}, \qquad \overline{U}_y = \frac{-3\overline{x}\overline{y}}{2(\overline{x}^2 + \overline{y}^2)^{5/2}} \tag{5.3}$$

where $\overline{U}_x = U_x/U_{x0}$, $\overline{U}_y = U_y/U_{x0}$, $\overline{x} = x/R$, and $\overline{y} = y/R$ (R is the radius of the sphere).

Table 5.1. The effect of gravity on the sedimentation factor of particles under conditions of potential heterogeneous flow past a sphere ($\varepsilon = \tau_{\mathrm{p}0}g/U_{x0}$)

Skt_f	η			η		
	downward flow			upward flow		
	$\varepsilon = 0.05$	$\varepsilon = 0.1$	$\varepsilon = 0.2$	$\varepsilon = 0.05$	$\varepsilon = 0.1$	$\varepsilon = 0.2$
0	0.0025	0.0081	0.029	0	0	0
0.3	0.16	0.21	0.29	0.048	0	0
0.7	0.40	0.44	0.50	0.54	0.49	0.31
2	0.67	0.71	0.74	0.62	0.58	0.49
5	–	–	–	–	0.79	0.74

The center of the coordinate system was located at the center of the sphere, and the x-axis was directed downstream, so that the critical point had the coordinates $\bar{x} = -1$ and $\bar{y} = 0$.

The Lagrangian equations of particle motion were written as

$$Stk_f\overline{V}_x\frac{\mathrm{d}\overline{V}_x}{\mathrm{d}\bar{x}} = \overline{U}_x - \overline{V}_x, \tag{5.4}$$

$$Stk_f\overline{V}_x\frac{\mathrm{d}\overline{V}_y}{\mathrm{d}\bar{x}} = \overline{U}_y - \overline{V}_y, \tag{5.5}$$

where $Stk_f = \tau_{\mathrm{p}0}U_{x0}/R$.

The boundary conditions for (5.4) and (5.5) are as follows: $\bar{x} = -\infty$, $\overline{V}_x = \overline{U}_x = 1$, and $\overline{V}_y = \overline{U}_y = 0$.

Michael and Norey [11] then changed from Cartesian to cylindrical coordinates.

The trajectories of particles motion were obtained as a result of computations. Figure 5.2 gives the limiting trajectories of particles (corresponding to $\bar{y}_{\mathrm{cr}}$) for different values of Stk_f. One can use the foregoing data to readily determine the values of the sedimentation factor of particles as a function of the Stokes number in averaged motion. These values are $\eta = 0.035, \eta = 0.35$, and $\eta = 0.82$ for the values of Stokes number $Stk_f = 0.2$, $Stk_f = 0.7$, and $Stk_f = 5$, respectively.

Michael and Norey [11] further investigated the effect of gravity on the sedimentation of particles. They treated two cases, namely, those of downward and upward flows of gas suspension past a sphere. For this purpose, the term $\varepsilon = \pm\tau_{\mathrm{p}0}g/U_{x0}$ (where g is the acceleration of gravity) was introduced into the right-hand part of (5.4). It is obvious that the effect of gravity will be significant only when the free-fall velocity of particles $\tau_{\mathrm{p}0}g$ and the velocity U_{x0} of flow in which they are suspended are of the same order of magnitude.

Table 5.1 gives values of the sedimentation factor of particles as a function of their inertia. One can see from the data in the table that the inclusion of

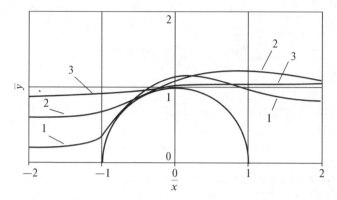

Fig. 5.2. Limiting trajectories of particles under conditions of potential heterogeneous flow past a sphere: (1) $Stk_f = 0.2$, (2) $Stk_f = 0.7$, (3) $Stk_f = 5.0$

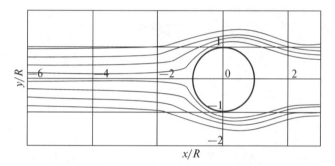

Fig. 5.3. Trajectories of particles $10\,\mu m$ in diameter ($Stk_f = 0.2$) under conditions of transverse potential heterogeneous flow past a cylinder: ($\rho_p = 1,400\,kg\,m^{-3}$, $U_{x0} = 6\,m\,s^{-1}$)

gravity causes an increase in the sedimentation factor in the case of downward flow past a sphere and a decrease in the sedimentation factor in the case of upward flow. Therefore, as the particle inertia increases and the flow velocity decreases, the failure to include the gravity force may lead to significant errors.

In a more recent study, Morsi and Alexander [12] calculated particle trajectories under conditions of transverse potential flow past a cylinder. In this study (as in [11]), the motion of single particles was treated, where interparticle collisions and their effect on gas were ignored. Compared to [11], an attempt was made to include the difference of the particle drag from that according to the Stokes law and particle motion from the Saffman lift force.

The particle trajectories obtained as a result of calculations are given in Figs. 5.3–5.5.

The studies cited above dealt with rather idealized cases of flow past bodies. The calculations did not include the effect of the viscous boundary layer developing on the body subjected to flow, and the motion of particles

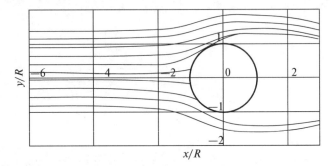

Fig. 5.4. Trajectories of particles $20\,\mu\mathrm{m}$ in diameter ($Stk_\mathrm{f} = 0.83$) under conditions of transverse potential heterogeneous flow past a cylinder: ($\rho_\mathrm{p} = 1,400\,\mathrm{kg\,m^{-3}}$, $U_{x0} = 6\,\mathrm{m\,s^{-1}}$)

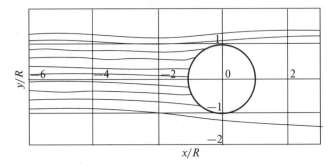

Fig. 5.5. Trajectories of particles $100\,\mu\mathrm{m}$ in diameter ($Stk_\mathrm{f} = 20$) under conditions of transverse potential heterogeneous flow past a cylinder: ($\rho_\mathrm{p} = 1,400\,\mathrm{kg\,m^{-3}}$, $U_{x0} = 6\,\mathrm{m\,s^{-1}}$)

reflected from the body was not treated, as well as the inverse effect of particles on gas. The nonisothermality of flow, which causes the emergence of the thermophoresis force, may also have a significant effect on the process of dust-laden flow past a body. In what follows, we will consider the results of studies whose authors tried to take into account some or the other of the physical factors listed above.

A viscous heterogeneous flow past the frontal surface of a sphere at values of the Reynolds number of $Re_\mathrm{pm} = U_{x0}R/\nu = 10^3 - 10^7$ was treated by Tsirkunov [24]. The carrier gas was assumed to be incompressible, and the concentrations of particles – negligible, so that the particles have no effect on the flow of continuous medium. The calculations revealed that the boundary layer distorts strongly the trajectory of particles and prevents them from moving toward the wall. This is attributed to the fact that the viscous gas is decelerated more intensively than the ideal one; this, in turn, leads to a more intensive deceleration of solid particles. The particles moving in the boundary layer in the vicinity of the surface subjected to flow lose their velocity abruptly, "hover," and then drift along the body surface. As a result, the sedimentation factor of particles decreases. Tsirkunov [24] further made an important inference that, at $Stk_f = \tau_{p0}U_{x0}/R \geq 0.2$, the boundary layer hardly affects the particle motion, while at $Stk_f \leq 0.11$ its effect is significant. The critical value of the Stokes number Stk_{fcr} corresponding to collisionless flow of particles past a body decreases with increasing Reynolds number Re_{pm}. This is apparently associated with the fact that an increase in the Reynolds number leads to a decrease in the boundary layer thickness, which causes a reduction of the forcing-back effect of the boundary layer.

The characteristic features of weakly dust-laden heterogeneous flow past a cylindrical surface under nonisothermal conditions were considered in the review of Spokoyny and Gorbis [23]. Analysis of the process of nonisothermal sedimentation revealed that, in the region of low-inertia particles for which the inertial mechanism of sedimentation is no longer valid ($Stk_f < Stk_{fcr}$), the intensity of sedimentation increases abruptly with nonisothermality and is largely defined by thermophoresis.

The conditions of inertial sedimentation of Stokesian particles in the case of a laminar heterogeneous jet flowing from a plane-parallel channel were studied by Dombrovsky and Yukina [6–8]. The Stokes number was determined as follows: $Stk_{fm} = \tau_{p0}|\partial U_x/\partial y|_{x,y=0}$. It was demonstrated that, at $Stk_{fm} > 0.5$, the decrease in the sedimentation factor due to the variation of trajectories of particles in the boundary layer does not exceed 15% even at low values of the Reynolds number. However, the components of velocity of particles at the wall suffered a significant decrease. This fact becomes important in studying the mechanical impact of particles on the wall surface. Dombrovsky and Yukina [8] further study the effect of the blowing of gas off the obstacle surface on the conditions of sedimentation. It has been demonstrated that, at high values of the Reynolds number, the blowing off of gas hardly affects the critical Stokes number in the investigated range of variation of the parameters.

In all of the studies referred to above, it was assumed that the particles which get to the body surface disappear from the flow. This formulation of the problem is acceptable when the dispersed phase is taken to be provided by liquid droplets or particles which form a thin film along the surface subjected to flow after getting to the body.

Vittal and Tabakoff [32] investigated a heterogeneous flow past a cylinder in view of the boundary layer, of the inverse effect of particles on gas, and of the effect of reflected particles. A flow with a relatively low-volume concentration of particles was treated; therefore, the interparticle interaction was disregarded. The parameters of continuous medium were calculated using the Eulerian approach, and the force of aerodynamic drag alone was taken into account in the Lagrangian equations of particle motion.

As a result of collisions with the surface, particles lose a part of their momentum and change the direction of motion. In order to calculate the particle trajectory after collision with the wall, one needs to know the magnitude and direction of the particle velocity vector. As is observed in [32], the rebound parameters are statistical and are largely defined by the particle angle of incidence. The following empirical relations were used in the calculations for the recovery factors of velocity after impact:

$$\frac{V_{n_2}}{V_{n_1}} = 1 - 0.4159\beta + 0.4994\beta^2 - 0.292\beta^3, \tag{5.6}$$

$$\frac{V_{\tau_2}}{V_{\tau_1}} = 1 - 2.12\beta + 3.0775\beta^2 - 1.1\beta^3, \tag{5.7}$$

where V_{n_1} and V_{τ_1} denote the normal and tangential (to the body surface) components of particle velocity prior to collision, respectively; and V_{n_2} and V_{τ_2} denote the components of particle velocity after collision, respectively. In (5.6) and (5.7), β is the angle (in radians) between the direction of particle velocity prior to collision and the tangent to the surface.

The thus calculated trajectories of quartz particles ($\rho_p = 2,444\,\mathrm{kg\,m^{-3}}$) of different sizes moving in an air flow past a cylinder ($R = 1.5675\,\mathrm{mm}$) for the Reynolds number $Re_{\mathrm{pm}} = 2U_{x0}R/\nu = 40$ are given in Figs. 5.6–5.8.

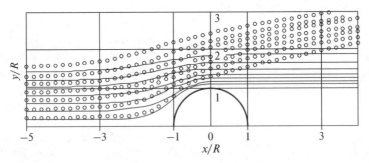

Fig. 5.6. Trajectories of particles $10\,\mu\mathrm{m}$ in diameter ($Stk_\mathrm{f} = 0.09$) under conditions of transverse heterogeneous flow past a cylinder: ($\rho_\mathrm{p} = 2,444\,\mathrm{kg\,m^{-3}}$, $Re_{\mathrm{pm}} = 40$): *lines* indicate potential flow, and *circles* indicate viscous flow

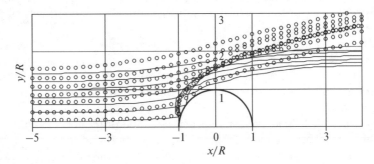

Fig. 5.7. Trajectories of particles $40\,\mu\mathrm{m}$ in diameter ($Stk_\mathrm{f} = 1.4$) under conditions of transverse heterogeneous flow past a cylinder: ($\rho_\mathrm{p} = 2,444\,\mathrm{kg\,m^{-3}}$, $Re_{\mathrm{pm}} = 40$): *lines* indicate potential flow, and *circles* indicate viscous flow

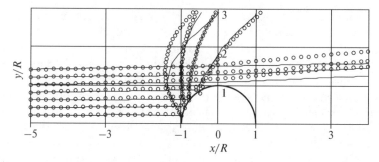

Fig. 5.8. Trajectories of particles 140 μm in diameter ($Stk_f = 17$) under conditions of transverse heterogeneous flow past a cylinder: ($\rho_p = 2,444\,\mathrm{kg\,m^{-3}}$, $Re_{pm} = 40$): *lines* indicate potential flow, and *circles* indicate viscous flow

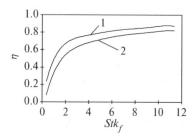

Fig. 5.9. The sedimentation factor of particles under conditions of flow past a cylinder 3.175 mm in diameter ($Re_{pm} = 40$) as a function of the Stokes number: (1) potential flow, (2) viscous flow

One can see that the trajectories of different particles differ strongly form one another in that the small particles do not collide with the body surface while the large particles collide with the body and deflect sideways. The particle motion is affected significantly by the boundary layer. The inclusion of the viscosity of gas results in an increase in the effective size of the cylinder by the displacement thickness of boundary layer; this affects the trajectories of particles and causes a decrease in their sedimentation factor. A graphic proof of the foregoing is provided by Fig. 5.9 which gives the distributions of the trapping coefficients of particles for the cases of both viscous and ideal fluid flow past a cylinder.

A disadvantage of the study [32] is the failure to include the force of gravity which, for the conditions of this investigation, should have made a significant impact by causing a variation of the trajectories of particles and disturbing the flow symmetry in the treated case of horizontal flow.

Some researchers investigated supersonic heterogeneous flow past bodies [21, 22]. However, these theoretical studies were performed for low values of mass concentration of particles, which enables one to study the dynamics of both incident and reflected particles in the preassigned velocity field of gas.

The simulation of external supersonic heterogeneous flow past blunt bodies
in view of the inverse effect of particles on gas was performed in [4, 5]. The
dynamics of particles were calculated using the Eulerian continuous approach.
Trajectories of particles under conditions of transverse dust-laden flow of a
plate of finite thickness were obtained in these studies. The distributions of
longitudinal velocity of gas in the presence of particles of different sizes and
of particles proper ignoring their reflection from the body surface, which were
obtained by Davydov and Nigmatulin [4], are given in Fig. 5.10. The resultant data indicate that the velocities of the gas and dispersed phases differ
significantly. The velocity of particles on the plate surface increases with the
particle inertia. The presence of large particles in the flow makes a strong
impact on the distribution of carrier gas velocity in the stagnation region.

Figures 5.11 and 5.12 give streamlines of carrier gas and suspended particles [4]. One can see that small particles tend to follow the lines of flow of

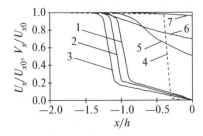

Fig. 5.10. Profiles of the longitudinal component of (1–3) the gas velocity in the
presence of particles and of (4–7) the velocity of particles, $M = 5, Ma_0 = 5.0$: (1, 5)
$d_p = 4.2\,\mu\text{m}$, (2, 6) $d_p = 42\,\mu\text{m}$, (3, 7) $d_p = 420\,\mu\text{m}$, (4) $d_p \to 0$

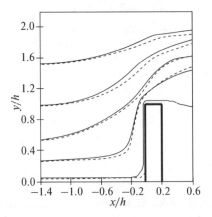

Fig. 5.11. Streamlines of carrier gas (*solid lines*) and small particles (*dotted lines*)
under conditions of heterogeneous flow past a plate: $d_p = 0.42\,\mu\text{m}$, $M = 5$,
$Ma_0 = 5.0$

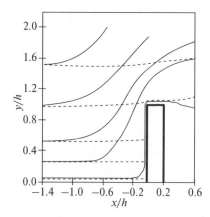

Fig. 5.12. Streamlines of carrier gas (*solid lines*) and large particles (*dotted lines*) under conditions of heterogeneous flow past a plate: $d_\mathrm{p} = 420\,\mu\mathrm{m}$, $M = 5$, $Ma_0 = 5.0$

gas, and their trajectories are curved significantly. On the other hand, large particles hardly change their almost straight trajectories in the vicinity of the plate surface.

Davydov et al. [5] calculated a flow past a plate in view of particles reflected from the frontal surface. The investigation was performed within the suggested mathematical model of three-velocity and three-temperature medium. Therefore, a "phase" of reflected particles which fly from the surface of the body subjected to flow toward the incident heterogeneous flow is introduced in the vicinity of the frontal surface along with the gas phase and "phase" of incident particles. At some values of concentration of the dispersed phase, collisions occur between incident and reflected particles; these collisions result in the variation of velocities of particles of both types. This causes the need for taking into account some effective force of interaction between particles from among the acting forces, as well as the "phase" transition of incident to reflected particles and vice versa. Because the resultant "phase" transition will be the transition from reflected to incident particles, only one source term is introduced in the continuity equations of these "phases" [5].

The following relations were employed in calculating the particle/wall interaction:

$$\frac{V_{n_2}}{V_{n_1}} = 1, \frac{V_{\tau_2}}{V_{\tau_1}} = 0.7. \tag{5.8}$$

Analysis of the set of equations obtained by Davydov et al. [5] revealed eight dimensionless parameters which define the intensity of physical processes under conditions of supersonic heterogeneous flow past bodies. These parameters include the Mach number of unperturbed flow, the adiabatic exponent for gas, the recovery factor of longitudinal component of velocity, the mass concentration of particles in unperturbed flow, the degree of inertia of particles, the parameter of velocity nonequilibrium of incident and reflected particles, the parameter which characterizes the variation of mass concentration due to collisions, and the Reynolds number calculated by the particle diameter.

5.2.2 Experimental Investigations

This section deals with the results of experimental investigation of the characteristics of heterogeneous flow in the neighborhood of the frontal point of the body subjected to flow [27, 28]. Cylinders with ends of different configurations were used as models. Primary consideration was given to the study of trajectories of solid particles in the vicinity of bodies subjected to flow and to the interaction between the dispersed phase and the surface of models.

The employed setup is described in Chap. 3. The experiments were performed for a downward turbulent flow of air in a pipe of inside diameter $D = 64\,\mathrm{mm}$. The Reynolds number was $Re_D = 11,200$ with the averaged velocity of air on the pipe axis $U_{xc} = 2.8\,\mathrm{m\,s^{-1}}$. The models (cylinders 11 mm in diameter) were placed within the pipe such that the cylinder axis and the pipe axis coincided (see Fig. 3.21). Spherical particles of glass of different sizes were used as the dispersed phase in the experiments.

The objective was to study the dynamics of large solid particles in the vicinity of the body subjected to flow. The measure of size or inertia of particles is characterized by their Stokes number in averaged motion. This Stokes number is determined as follows (see Table 1.1):

$$Stk_f = \frac{\tau_p}{T_f}, \tag{5.9}$$

where τ_p is the time of dynamic relaxation of particle, and T_f is the characteristic time of carrier gas in averaged motion. The characteristic time may be estimated as

$$T_f = \frac{L}{U_{xc}}, \tag{5.10}$$

where L is the distance from the critical point of the body upstream, at which the curvature of streamlines of gas begins. In a first approximation, one can assume $L \approx R$, where R is the cylinder radius.

Note that the characteristic time of the carrier phase in averaged motion is determined in a different manner from the case of heterogeneous flows in channels (see Sect. 3.6). The characteristic time of the carrier phase in averaged motion in the neighborhood of the frontal point of a model becomes much shorter. As a result, the weakly dust-laden nonequilibrium heterogeneous flow incident on a body in the vicinity of its surface may be classified as a flow with large particles. Evidence of this may be provided by the calculated values (given in Table 5.2) of the Stokes number of particles used in the experiments under the conditions described above.

Table 5.2. Characteristics of employed spherical particles

no.	material	rated diameter, μm	density of particle material	Stokes number in averaged motion (experiment), Stk_f	Stokes number in averaged motion (calculation), Stk_f
1	SiO_2	50	2,550	10.5	9
2	SiO_2	100	2,550	32	29
3	SiO_2	200	2,550	82	77

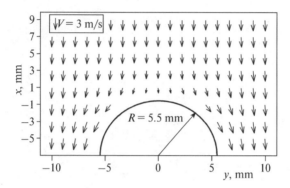

Fig. 5.13. The velocity field for air in the neighborhood of a cylinder with a hemispherical end

A large particle is a particle whose Stokes number is of the order of ten or higher. In this case, particles do not follow the streamlines of carrier gas, which bend in the vicinity of the body, deviate from these lines, and experience collisions with the body. The emergence of the "phase" of reflected particles in the flow complicates the flow pattern significantly.

We will now analyze the obtained results. Figures 5.13–5.15 give the velocity fields for air and particles incident on and reflected from the body. The mass concentration of particles during measurements was $M = 0.007$. As was demonstrated by the result of investigations described in Chap. 4, the presence in the flow of particles in such insignificant concentrations does not affect the parameters of flow of the carrier phase. Indeed, the value of volume concentration of particles $\Phi = 3.3 \times 10^{-6}$ (at which the mode with single particles is realized, see Fig. 1.8) corresponds to this value of mass concentration of the dispersed phase.

The obtained results indicate that the deceleration of air (curvature of the lines of flow) begins from a distance $x \approx R$ (see Fig. 5.13). The data of Fig. 5.14 clearly indicate that the sedimentation factor of particles employed in the experiment was close to unity, $\eta \approx 1$.

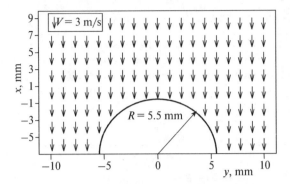

Fig. 5.14. The velocity field for incident glass particles (100 μm) in the neighborhood of a cylinder with a hemispherical end

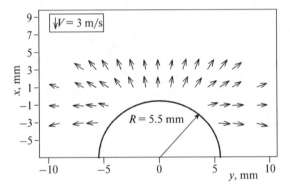

Fig. 5.15. The velocity field for glass particles (100 μm) reflected from the surface of a cylinder with a hemispherical end

The velocity field for solid particles given in Fig. 5.15 in the immediate vicinity of the model surface may be used to determine the recovery factor of the longitudinal and normal components of velocity of particles in their interaction with the wall. For example, the recovery factor of velocity of particles incident on the critical point of a body is $V_{x2}/V_{x1} \approx 0.8$.

Note that the problem of construction of trajectories of solid particles in the neighborhood of the critical point of a body using the measurement results is far from simple. The presence at some arbitrary points of space of some characteristic distributions of velocities corresponding to particles of different "types" seriously complicates analysis of the experimental data. Indicative of this are the results given below.

The scheme of a flow of gas with large solid particles is given in Fig. 5.16. Examples of measured distributions of velocities of the dispersed phase at some selected points are given in Fig. 5.17.

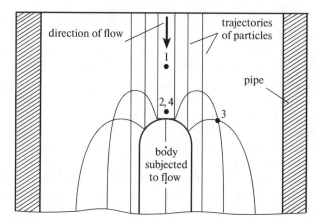

Fig. 5.16. The scheme of a heterogeneous flow in the vicinity of a cylinder with a hemispherical end. The measured distributions of particle velocities at characteristic points 1, 2, 3, and 4 given in Figs. 5.17a, 5.17b, 5.17c, and 5.17d, respectively

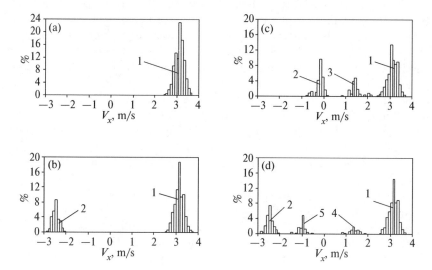

Fig. 5.17. Measured distributions of velocities of glass particles ($100\,\mu$m): (a) $x = 10\,$mm, $y = 0, M = 0.007$; (b) 1, 0, 0.007; (c) -1, 6, 0.007; (d) 1, 0, 0.4. The numerals indicate the distributions of (1) incident particles, (2) reflected particles, (3) reflected particles with "positive" velocities, (4) incident particles after collision with reflected ones, (5) reflected particles after collision with incident ones

The center of the rectangular coordinate system $(x - y)$ is located at the critical point of the body. The x-axis is directed upstream. The location of measurement points 1–4 is shown schematically in Fig. 5.16.

The distribution of particle velocities in an unperturbed flow on the pipe axis at a distance $x = 10\,$mm from the critical point of the body (point 1) is given in Fig. 5.17a. One can see that the averaged velocity of incident particles in the flow unperturbed by the presence of the body is $V_x \approx 3.1\,\mathrm{m\,s^{-1}}$, i.e., exceeds the carrier air velocity, which is not surprising in the case of downward flow. One can further infer from the data of Fig. 5.17a that the mean-square deviation of particle velocity is $\sigma_{V_x} \approx 10\%$. The numerous reasons for the emergence of fluctuations of velocities of solid particles suspended in a turbulent gas flow are analyzed in Chap. 4.

The distribution of particle velocities at point 2 (see Fig. 5.16) located at a distance $x = 1\,\mathrm{mm}$ from the body surface at $M = 0.007$ is given in Fig. 5.17b. One can see that a "phase" of reflected particles arises in addition to that of incident particles, with the reflected particles moving toward the main flow; therefore, their velocities assume negative values. The averaged velocity of reflected particles is lower than that of incident particles and is $V_x \approx 2.5\,\mathrm{m\,s^{-1}}$.

The measured velocities of particles at point 3 ($x = -1\,\mathrm{mm}$, $y = 6\,\mathrm{mm}$) for a low concentration of the dispersed phase are given in Fig. 5.17c. The data are clearly indicative of the presence of distributions of three characteristic types. The distribution of the first type relates to incident particles which experienced no collisions with the body. The averaged velocity of these particles is $V_x \approx 3.1\,\mathrm{m\,s^{-1}}$. The distribution of the second type relates to reflected particles for which the value of the recovery factor of the longitudinal component of velocity vector is low, $V_x \approx 0.2\,\mathrm{m\,s^{-1}}$. These particles change the direction of their motion by almost 90° after collision with the body. The distribution of the third type corresponds to reflected particles with "positive" velocities, $V_x = 1.4\,\mathrm{m\,s^{-1}}$. These particles, which moved toward the main flow after rebounding from the body, stopped and began accelerating again along the body surface. The trajectories of such particles are shown schematically in Fig. 5.16.

The measured particle velocities at point 4 (see Fig. 5.16) located at a distance $x = 1\,\mathrm{mm}$ from the body surface for a high concentration of the dispersed phase in incident flow ($M = 0.4$) are given in Fig. 5.17d. One can see that, compared to the case of low concentration of particles (see Fig. 5.17b), two new characteristic distributions of velocity appear. The probability of collision of incident and reflected particles increases with the concentration of particles. It is apparently with this factor that two experimentally observed "additional" distributions of velocities are associated, which relate to colliding incident particles ($V_x \approx 1.4\,\mathrm{m\,s^{-1}}$) and reflected particles ($V_x \approx 1\,\mathrm{m\,s^{-1}}$).

The following point is important. As the concentration of particles increases, a tendency is observed for the convergence of velocities of incident and reflected particles and for the increase in the probability of repeated collisions between them; this leads to entanglement of the particle trajectories. As a result, the obtained distributions of velocities will not exhibit clearly defined maxima but will be more "diffuse."

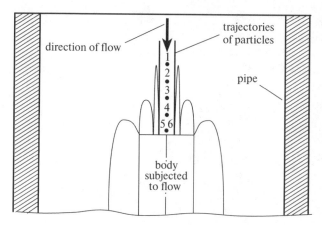

Fig. 5.18. The scheme of a heterogeneous flow in the vicinity of a cylinder with a flat end. The measured distributions of velocities of particles 100 μm in diameter at characteristic points 1, 2, 3, 4, 5, and 6 are given in Figs. 5.20a, 5.20b, 5.20c, 5.20d, 5.20e, and 5.20f, respectively

Described below are the results of investigation of the dynamics of large solid particles under conditions of longitudinal heterogeneous flow of a cylinder 11 mm in diameter with a flat end. The scheme of gas flow for this case is given in Fig. 5.18. Examples of measured distributions of velocities of the dispersed phase at some selected points are given in Figs. 5.19–5.21.

The distributions of velocities of glass particles ($d_p = 50 \, \mu m$) are given in Fig. 5.19. The distribution of the first type relates to incident particles. The averaged velocity of these particles is $V_x \approx 3 \, \mathrm{m \, s^{-1}}$. The distribution of the second type corresponds to reflected particles which move toward the main flow. The velocity of these particles in the vicinity of the wall ($x = 1 \, mm$) is $V_x \approx 2.4 \, \mathrm{m \, s^{-1}}$ (see Fig. 5.19f). The velocity of reflected particles decreases with increasing distance from the wall. The "phase" of reflected particles disappears at a distance $x \approx 15 \, mm$ from the wall (see Fig. 5.19a). The distribution of the third type relates to incident particles which made a transition from the "phase" of reflected particles. These particles, on the contrary, accelerate on approaching the body, and their velocity at a distance $x = 1 \, mm$ from the model is $V_x \approx 1.3 \, \mathrm{m \, s^{-1}}$ (see Fig. 5.19f).

The distributions of velocities of larger glass particles ($d_p = 100 \, \mu m$) are given in Fig. 5.20. The averaged velocity of incident particles is $V_x \approx 3.2 \, \mathrm{m \, s^{-1}}$, which exceeds the velocity of smaller particles with a rated diameter of 50 μm. The measurement results given in Fig. 5.20 enable one to analyze the dynamics of particles reflected from the model surface and of incident particles which experienced a collision with the body and began accelerating again while moving toward the model. The distribution indicated by numeral 4 corresponds

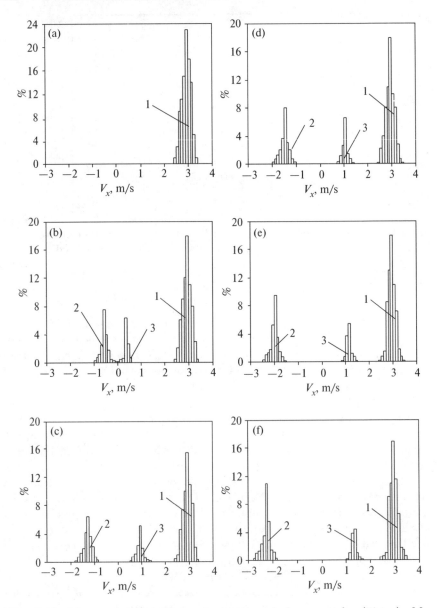

Fig. 5.19. Measured distributions of velocities of glass particles (50 μm), $M = 0.007$, $y = 0$: (a) $x = 15$ mm, (b) 12, (c) 9, (d) 6, (e) 3, (f) 1. The numerals indicate the distributions of (1) incident particles, (2) reflected particles, (3) incident particles after collision with the model surface

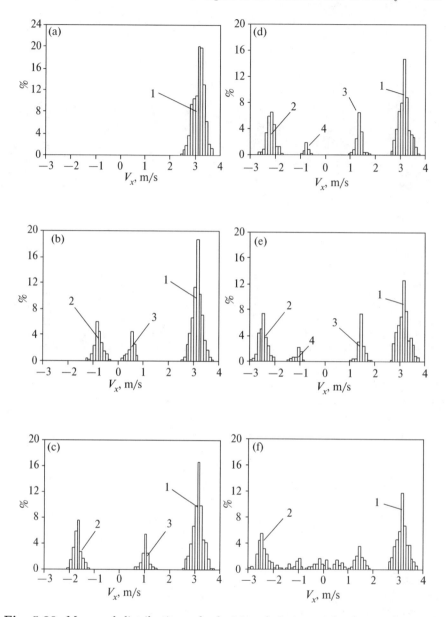

Fig. 5.20. Measured distributions of velocities of glass particles (100 μm), $y = 0$:
(a) $x = 40$ mm, $M = 0.007$; (b) 30, 0.007; (c) 20, 0.007; (d) 10, 0.007; (e) 1,
0.007; (f) 1, 0.4. The numerals indicate the distributions of (1) incident particles,
(2) reflected particles, (3) incident particles after the first collision with the surface,
(4) reflected particles after the second collision with the surface

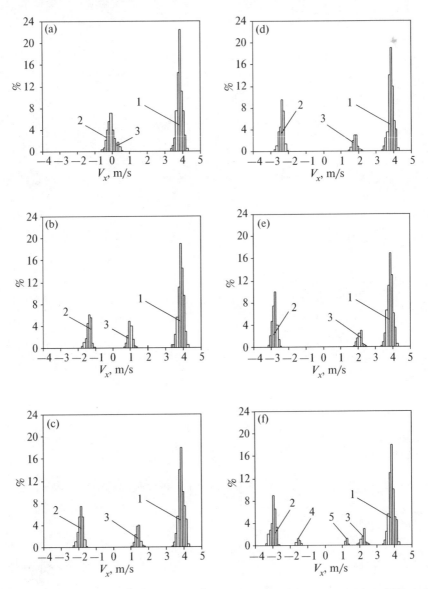

Fig. 5.21. Measured distributions of velocities of glass particles (200 μm), $M = 0.007, y = 0$: (**a**) $x = 100$ mm, (**b**) 80, (**c**) 60, (**d**) 40, (**e**) 20, (**f**) 1. The numerals indicate the distributions of (1) incident particles, (2) reflected particles, (3) incident particles after collision with the model surface, (4) reflected particles after the second collision with the model surface, (5) incident particles after the second collision with the model surface

to reflected particles after the second collision with the model. An increase in the concentration of particles leads to collisions between particles representing different "phases" such as incident, reflected, and so on. As a result, the measured distributions of velocities become less "clearly defined." For example, particles appear in the region of the critical point of the body, whose velocity is close to zero (see Fig. 5.20f). This fact causes an increase in the particle concentration in this region of flow. Note that large particles exhibit a higher inertia and, as a result, propagate through a larger distance toward the flow. Indeed, the rebound of large particles ($d_p = 100\,\mu\mathrm{m}$) is $x \approx 30\,\mathrm{mm}$, while the respective quantity for small particles ($d_p = 50\,\mu\mathrm{m}$) is $x \approx 12\,\mathrm{mm}$ (see Fig. 5.19).

Figure 5.21 gives the distributions of velocities of even larger particles of glass ($d_p = 200\,\mu\mathrm{m}$). The averaged velocity of incident particles is $V_x \approx 3.9\,\mathrm{m\,s^{-1}}$. Particles reflected from the surface have a velocity $V_x \approx 3\,\mathrm{m\,s^{-1}}$ (at a distance $x = 1\,\mathrm{mm}$). Because large particles are characterized by a high inertia, their deceleration after interaction with the model during the motion toward the flow is less intensive. At a distance $x \approx 100\,\mathrm{mm}$, the "phase" of reflected particles changes to the "phase" of incident particles. The respective distributions of velocities of particles "merge" (see Fig. 5.21a). The particles accelerate repeatedly and, as they approach the body, they reach a velocity $V_x \approx 2.1\,\mathrm{m\,s^{-1}}$ in the vicinity of the model. The distribution of velocities of particles reflected after the second collision with the surface at a distance $x = 1\,\mathrm{mm}$ are given in Fig. 5.21f. Also given in this figure is the distribution of velocities of incident particles which experienced two collisions with the body.

The distributions of velocities of glass particles with diameters $d_p = 50, 100$, and $200\,\mu\mathrm{m}$ as functions of the distance to the critical point of the body are given in Figs. 5.22–5.24, respectively. The data in the figures give a clear idea of the motion of glass particles of different sizes, including incident, reflected from the model, and repeatedly incident on the surface.

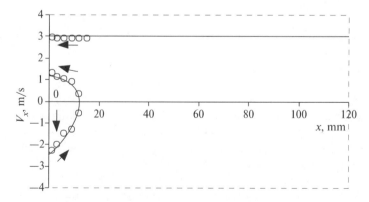

Fig. 5.22. The distribution of the velocity of glass particles ($50\,\mu\mathrm{m}$) in the neighborhood of a body with a flat end ($y = 0$)

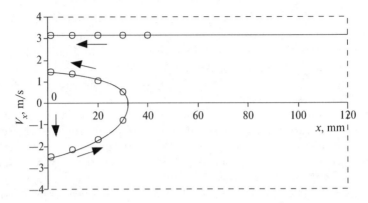

Fig. 5.23. The distribution of the velocity of glass particles (100 μm) in the neighborhood of a body with a flat end ($y = 0$)

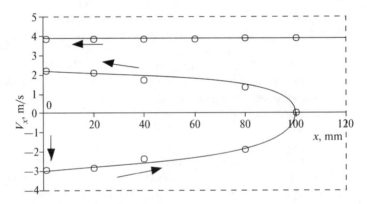

Fig. 5.24. The distribution of the velocity of glass particles (200 μm) in the neighborhood of a body with a flat end ($y = 0$)

We will systematize the experimental data described above for the purpose of determining the size of the region of existence of rebounded particles and of the value of the recovery factor of velocity of a particle reflected from the surface when it is repeatedly incident on the body subjected to flow.

Figure 5.25 gives the results of generalization of data on the rebound of particles as a function of their inertia (Stokes number). One can see in this figure that the rebound of particles from the model surface is directly proportional to the inertia of the dispersed phase in the investigated range of values of the Stokes number.

The effect of the inertia of the dispersed phase on the recovery factor of velocity of particles in their repeated interaction with the model surface (the ratio of velocities of incident particles on the wall in the secondary and primary collisions $k_{w31} = V_{x3}/V_{x1}$) is shown in Fig. 5.26. The data in the

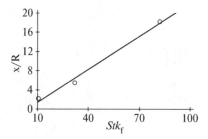

Fig. 5.25. The effect of the inertia of particles on the extent of their rebound

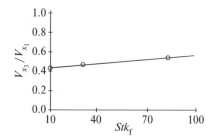

Fig. 5.26. The effect of the inertia of particles on the recovery factor of their velocities

figure indicate that this factor exhibits some tendency for an increase with increasing inertia of the dispersed phase and, for the described experimental conditions, is approximately $k_{w31} \approx 0.5$.

Now, a few words about the generalization of data on the sedimentation factor of particles. Analysis of the dependence of the sedimentation factor of particles on the Stokes number $\eta = \eta\,(Stk_f)$ reveals [10] that each distribution has two bends. These bends correspond to two "critical" values of the Stokes number, namely, minimal $Stk_{f\,min}$ and maximal $Stk_{f\,max}$. For low values of the Stokes number $(Stk_f < Stk_{f\,min})$, the sedimentation factor is negligible, i.e., $\eta \to 0$. In the other limiting case, that of high values of the Stokes number $(Stk_f > Stk_{f\,max})$, the sedimentation factor is close to unity, i.e., $\eta \to 1$. As to the order of magnitude of the given "critical" values of the Stokes number, it is clear from simple physical considerations that $Stk_{f\,min} = O(0.1)$ and $Stk_{f\,max} = O(10)$. Nevertheless, the presently available calculation data and experimental results do not enable one either to construct a "universal" curve of sedimentation of particles or to plot a family of curves for a fairly wide range of variation of flow parameters (determining dimensionless complexes). The region of likely values of the sedimentation (trapping) factor of particles given in Fig. 5.27 as a function of the most important parameter, i.e., the Stokes number, describes the majority of the investigation results given above.

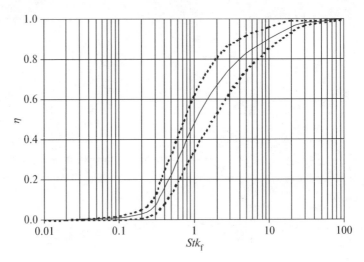

Fig. 5.27. The sedimentation factor of particles as a function of the Stokes number

Proceeding from the calculation data and experimental results described above, we will list the main factors which have an effect on the value of η:

1. The potentiality of flow (the effect of the boundary layer causes a decrease in η and becomes significant at $Stk_f < 0.2$).
2. The gravity force (causes a significant increase in η at $Stk_f < 0.3$ and $\varepsilon > 0.05$ in the case of downward flow and a significant decrease in η at $Stk_f < 2$ and $\varepsilon > 0.05$ in the case of upward flow).
3. The nonisothermality of flow (causes an increase in η at $Stk_f < 0.1$).
4. The axial symmetry of flow (the curvature of the streamlines of flow begins at a shorter distance compared to the case of plane flow, which causes an increase in the inertia of particles and in η).

5.3 A Particle-Laden Flow in the Boundary Layer of a Body Subjected to Flow

In this section, we analyze the results of theoretical and experimental investigations of the behavior of particles and their inverse effect on the parameters of gas flow in a boundary layer. The study of the effect of particles on a boundary-layer flow is a problem which is far from ordinary. In the light of the presently available experimental data, it appears obvious that the dispersed phase may have a dual effect on a near-wall flow. Firstly, the dispersed phase may affect the flow in a boundary layer by way of modification of incident flow. Secondly, particles make a direct impact on the flow in a boundary layer due to their inertia, namely, due to the presence of dynamic and thermal (in the case of nonisothermal flow) slip.

5.3.1 Theoretical Investigations

We will begin treatment from the simplest cases of flow. Osiptsov [15] studied the development of profiles of the longitudinal and transverse components of velocity of gas and solid particles, as well as of the concentration of particles in a laminar boundary layer of a semiinfinite flat plate. The theoretical investigation was performed within the model of two interpenetrating continua [14]. It was assumed that the particles are spheres of the same radii, and that their volume concentration was low. Because the physical density of particles is several orders of magnitude higher than the carrier gas density, the Stokes force was taken to be the only force of interphase interaction in the entire computational domain in the boundary layer.

The laminar heterogeneous flow treated by Osiptsov [15] may be classified (see Chap. 1) as quasiequilibrium, because the velocities of the carrier and dispersed phases were taken to be equal to each other. The presence of dynamic slip due to the inertia of particles in the flow in a boundary layer caused this flow to become nonequilibrium. In the calculations, the mass concentration of particles was varied in a wide range; this resulted in a significant impact of the dispersed phase on the parameters of gas flow.

Figure 5.28 gives the distributions of the longitudinal component of velocity for both phases of heterogeneous flow in a boundary layer, as well as the Blasius profile corresponding to laminar single-phase flow, for different values of dimensionless longitudinal coordinate. The latter coordinate was rendered dimensionless as follows:

$$\bar{x} = \frac{x}{l_p} = \frac{x}{\tau_p V_{x0}} = \frac{18x\mu}{\rho_p d_p^2 V_{x0}}, \tag{5.11}$$

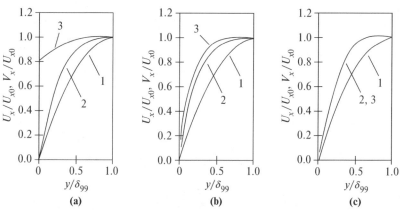

Fig. 5.28. The development of profiles of the longitudinal component of velocity of gas and particles ($M = 3$): (**a**) $\bar{x} = 0.2$ ($Stk_f = 5$), (**b**) $\bar{x} = 1.0$ ($Stk_f = 1$), (**c**) $\bar{x} = 12$ ($Stk_f = 0.083$); (1) Blasius profile, (2) carrier phase, (3) particles

where x is the longitudinal coordinate reckoned from the beginning of the plate, l_p is the length of dynamic relaxation (deceleration) of a particle, and V_{x0} is the velocity of particle in incident flow (in this case, it is equal to the respective velocity for gas, $V_{x0} = U_{x0}$).

The dimensionless length of dynamic relaxation $\bar{x}$ is the reciprocal of the local Stokes number in averaged motion (see Table 1.1), i.e., $Stk_f = 1/\bar{x}$. Note that this Stokes number is different from the respective Stokes numbers which characterize the processes of relaxation of averaged velocities of gas and particles under conditions of pipe flow (see Sect. 3.6) and in the neighborhood of the critical point of the body subjected to flow (see Sect. 5.2.1).

Figure 5.28 demonstrates that the longitudinal velocity of particles is higher than the gas velocity in the entire boundary layer; in so doing, at $\bar{x} < 1$ ($Stk_f > 1$) it is other than zero on the plate surface. This fact is due to the inertia of particles. The velocity difference between the phases leads to an intensive exchange of momentum; as a result, the velocity profile of the gas phase is much flatter compared to the case of single-phase flow. Osiptsov [15] observes that the relaxation of the phase velocities terminates in fact at $\bar{x} = 5$ ($Stk_f = 0.2$), and the flow structure is of the same type for different values of mass concentration of particles M. The profiles of longitudinal velocities of both phases at high values of $\bar{x}$ (low values of Stk_f) become self-similar. These limiting profiles may be obtained from the solution of Prandtl equations for a single-phase gas of higher density $\rho_e = \rho + \Phi\rho_p = \rho(1 + M)$. Therefore, after the relaxation of velocities, this flow once again (as in the case of flow incident on a plate) becomes quasiequilibrium.

Figure 5.29 gives profiles of the transverse component of velocity of the gas phase and particles. One can see in the figure that a region with $V_y < U_y$ exists in the boundary layer in the case of low values of $\bar{x}$ (high values of

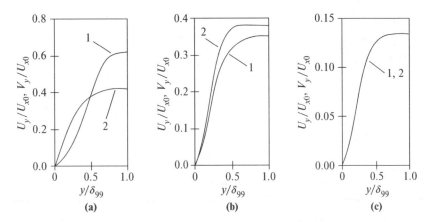

Fig. 5.29. The development of profiles of the transverse component of velocity of gas and particles ($M = 3$): (a) $\bar{x} = 0.2$ ($Stk_f = 5$), (b) $\bar{x} = 1.0$ ($Stk_f = 1$), (c) $\bar{x} = 12$ ($Stk_f = 0.083$); (1) carrier phase, (2) particles

Stk_f); this implies the crossing of the streamlines of carrier gas by particles in the wall direction. Osiptsov [15] further observes that the relaxation of the transverse components of phase velocities occurs over a much greater length than in the case of longitudinal components.

The obtained distributions of the concentration of particles demonstrated that, at $\bar{x} < 1$ ($Stk_f > 1$), the density of the dispersed phase increases monotonically on approaching the plate and reaches a finite value $M_w = M_0/(1 - \bar{x})$ on the wall. At $\bar{x} \geq 1$ ($Stk_f \leq 1$), the concentration of particles tends to infinity as the wall is approached.

More recently, Osiptsov [16] studied zones of unrestricted increase of concentration of particles in flows. Unfortunately, his inferences made in [16] were not supported by experimental results. Apparently, no such data are available at present either.

Osiptsov [15] further calculated local coefficients of friction c_f for different values of the mass concentration of particles. The distributions of this characteristic of flow are given in Fig. 5.30. The quantity $c_f\sqrt{Re_x}$ is plotted on the ordinate and is determined as

$$c_f\sqrt{Re_x} = \frac{\tau_w\sqrt{Re_x}}{\rho U_{x0}^2} = \mu\left(\frac{\partial U_x}{\partial y}\right)_w \frac{\sqrt{Re_x}}{\rho U_{x0}^2}, \qquad (5.12)$$

where τ_w is the local shear stress on the wall, and $Re_x = (U_{x0}x)/\nu$ is the local Reynolds number.

One can see in Fig. 5.30 that the value of coefficient of friction in a heterogeneous flow is much higher than in a single-phase flow. The friction increase on the wall at $\bar{x} < 1$ ($Stk_f > 1$) is attributed to the increase in the velocity gradient of carrier gas in this region of flow because of the interphase exchange of momentum. The velocity of particles downstream decreases and their concentration increases. As a result, the gas velocity gradient decreases to cause

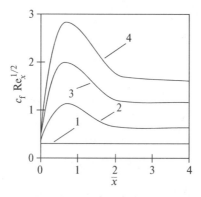

Fig. 5.30. The distribution of the coefficient of friction along a plate: (1) $M = 0$, (2) $M = 3$, (3) $M = 10$, (4) $M = 20$

a decrease in the coefficient of friction. At $\bar{x} \to \infty$ ($Stk_f \to 0$), the coefficient of friction tends to its limiting value

$$c_f = \frac{0.332(1+M)^{1/2}}{\sqrt{Re_x}}, \tag{5.13}$$

which corresponds to the Blasius solution for a single-phase gas of higher density.

It will be recalled that Osiptsov [15] used the Stokes force as the only force defining the motion of particles in a laminar boundary layer. Attempts at taking into account the effect of the Saffman lift force caused by the nonuniformity of the gas velocity field were made by Osiptsov [17] and Naumov [13]. As to another force which also causes transverse migration of particles, namely, the Magnus force, Naumov [13] showed that, at $Re_{pm} < 10$, where $Re_{pm} = U_{x0}d_p/\nu$, the projection of this latter force onto the transverse axis is much smaller than that of the Saffman force. Note that the Reynolds number Re_{pm} employed here in analyzing the flow in a laminar boundary layer is an analog of the particle Reynolds number Re_p constructed by the relative velocity between the phases, because the gas velocity in the near-wall region is close to zero, and the velocity of particles at low values of $\bar{x}$ (high values of Stk_f) differs little from the gas velocity in the external flow.

Figure 5.31 gives the distribution of the transverse velocity of particles along a plate [13]. Plotted on the abscissa are values of the dimensionless transverse velocity of particles $\bar{V}_y = V_y\sqrt{x/U_{x0}\nu}$. The prolate boundary-layer coordinate $\varphi = y\sqrt{U_{x0}/\nu x}$ is plotted on the ordinate. One can see that, at $\bar{x} < 2(Stk_f > 0.5)$, the transverse velocity of particles assumes negative values in the vicinity of the wall, i.e., it is directed toward the wall. This is attributed to the fact that, along with high gradients of longitudinal velocity of gas, a significant difference between the longitudinal velocities of the phases (see

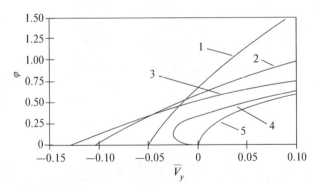

Fig. 5.31. The development of profiles of the transverse component of velocity of particles in the near-wall region ($M = 3$, $\bar{\rho} = 10^3$, $Re_{pm} = 1$): (1) $\bar{x} = 0.2$ ($Stk_f = 5$), (2) $\bar{x} = 0.4$ ($Stk_f = 2.5$), (3) $\bar{x} = 1.0$ ($Stk_f = 1$), (4) $\bar{x} = 2.0$ ($Stk_f = 0.5$), (5) $\bar{x} = 3.0$ ($Stk_f = 0.33$)

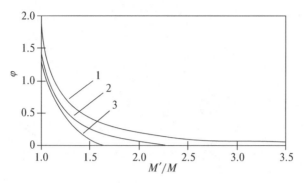

Fig. 5.32. The profiles of the concentration of particles at $M = 3, \bar{\rho} = 10^3, \bar{x} =$ 1.0, and $Stk_f = 1$: (1) $Re_{pm} = 0$, (2) $Re_{pm} = 1$, (3) $Re_{pm} = 2$. M' is the local concentration

Fig. 5.28a) is present in the near-wall region in the vicinity of the leading edge of the plate; this results in the emergence of the Saffman force. The longitudinal velocity of particles in this region of flow exceeds the gas velocity; therefore, the Saffman force acting on the particles causes their displacement in the plate direction. The difference between the longitudinal velocities of the phases decreases with increasing distance from the leading edge of the plate (see Fig. 5.28); this causes a decrease in the Saffman force. As a result, when $\bar{x}$ increases (Stk_f decreases), the region of negative values of the transverse velocity of particles narrows down and, at $\bar{x} = 3$ ($Stk_f = 0.33$), disappears.

Figure 5.32 gives the distributions of the concentration of particles in a single cross-section of the flow for different values of the Reynolds number Re_{pm} [13]. The distribution of the concentration of particles at $Re_{pm} = 0$, i.e., in the absence of dynamic slip between the phases, repeats exactly the distribution obtained by Osiptsov [15] disregarding the Saffman force. An increase in the Reynolds number causes an increase in the Saffman force; this, in turn, leads to an increase in the transverse velocities of particles. The flow of particles settling out on the wall causes a decrease in the concentration of impurity in the near-wall region. Therefore, an increase in the impact made by the Saffman force on the dynamics of particles is accompanied by a qualitative transformation of the profile of their concentration, which results in the disappearance of the region with the maximum of concentration of particles in the vicinity of the wall. A similar inference was made in [17] in calculating the motion of dust-laden gas in the initial region of a flat channel and round pipe, where the effect of the Saffman force is also significant.

The inclusion of the Saffman force leads to a significant variation of the distributions of the coefficient of friction along the plate in a laminar boundary layer. Figure 5.33 gives calculated values of the coefficient of friction for different values of the Reynolds number Re_{pm} [13]. The dashed line corresponds to the Blasius solution for a single-phase laminar boundary layer.

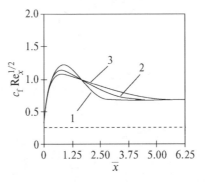

Fig. 5.33. The distribution of the coefficient of friction along a plate ($M = 3, \bar\rho = 10^3$): (1) $Re_{\mathrm{pm}} = 0$, (2) $Re_{\mathrm{pm}} = 1$, (3) $Re_{\mathrm{pm}} = 2$

The distribution of friction on a plate at $Re_{\mathrm{pm}} = 0$ hardly differs from that obtained by Osiptsov [15] and given in Fig. 5.30. As the Reynolds number increases, the concentration of particles in the wall region decreases (as was shown above); this causes a decrease in the intensity of interphase exchange of momentum. As a result, the gas velocity profile is less flat, the velocity gradient on the wall decreases, and the maximal value in the distribution $c_f\sqrt{Re_x}$ decreases as well (see Fig. 5.33). At the same time, the decrease in the coefficient of friction along the plate at the Reynolds number other than zero is smoother. This smoothness is attributed to the fact that the presence of particles moving in the wall direction (these particles are characterized by higher value of longitudinal velocity than the particles moving in the immediate vicinity of the wall) causes an increase in the extent of the region of relaxation of longitudinal velocities of the phases, in which high values of the coefficient of friction are observed.

The foregoing leads one to infer that the failure to include the Saffman lift force acting on particles in the calculation of a laminar heterogeneous boundary layer at $Re_{\mathrm{pm}} \leq 1$ may result in significant errors.

5.3.2 Experimental Investigations

The characteristics of a turbulent heterogeneous boundary layer developing on a flat plate were studied by Rogers and Eaton [18–20]. The distributions of averaged velocities of air and particles of glass are given in Fig. 5.34. The data in the figure show that the particles in upward flow move slower than air. The difference between the velocities of the dispersed and gas phases is close to the free-fall velocity and is almost constant over the entire boundary layer. The cross-section being treated is located at a distance $x = 0.55$ m from the beginning of the plate, which corresponds to the values of the Stokes number $Stk_f = 0.24$ and $Stk_f = 0.63$ for particles 50 and 90 μm in size, respectively. These values of the Stokes number indicate that the relaxation of velocities of

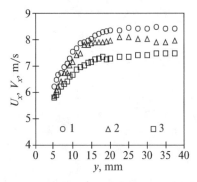

Fig. 5.34. The distributions of averaged velocities in a turbulent boundary layer on a plate ($x = 0.55$ m, $U_{x0} = 8$ m s^{-1}, $Re_x = 2.9 \times 10^5$, $M = 0.02$): (1) air, (2) glass particles (50 μm), (3) glass particles (90 μm)

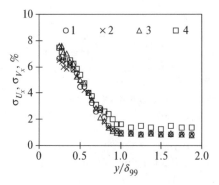

Fig. 5.35. The distributions of the longitudinal component of fluctuation velocity in a turbulent boundary layer ($x = 0.55$ m, $U_{x0} = 8$ m s^{-1}, $Re_x = 2.9 \times 10^5$): (1) air, ($M = 0$), (2) air ($M = 0.02$), (3) glass particles (50 μm), (4) glass particles (90 μm)

the phases has in fact terminated on reaching this cross-section. The experimental results demonstrated that, at $M = 0.02$, the particles made no impact on the distribution of averaged velocity of carrier air.

Profiles of the longitudinal component of fluctuation velocity of "pure" air, of air in the presence of particles, and of glass particles of different sizes in a turbulent boundary layer are given in Fig. 5.35. One can infer that the presence of particles in the flow had no effect on the distribution of fluctuation velocity of the carrier phase in the boundary layer. The magnitude of fluctuations of velocities of glass particles 50 μm in diameter is close to the respective characteristic for air. The fluctuations of velocities of large particles 90 μm in diameter exceed those of the carrier phase. Simple estimates indicate that the smaller particles must be readily involved in the fluctuating motion by turbulent eddies. As to the larger particles, the high values of fluctuation velocities are due to the use of polydisperse particles in the experiments

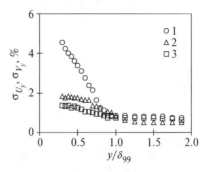

Fig. 5.36. The distribution of the transverse component of fluctuation velocity in a turbulent boundary layer ($x = 0.55\,\mathrm{m}$, $U_{x0} = 8\,\mathrm{m\,s}^{-1}$, $Re_x = 2.9 \times 10^5$, $M = 0.02$): (1) air ($M = 0$), (2) glass particles ($50\,\mu\mathrm{m}$), (3) glass particles ($90\,\mu\mathrm{m}$)

(see Sect. 4.2.2). The increase in fluctuation velocities of particles of both types in the wall region (where the relative inertia of the dispersed phase increases with decreasing characteristic times of energy-carrying eddies) is attributed to the nonuniformity of distributions of averaged velocities of particles.

Figure 5.36 gives the distribution of the normal component of fluctuation velocity. One can infer that the particle fluctuation velocities in the direction being treated are lower than the respective fluctuation velocities of air. The difference between the fluctuations of velocities of the gas and dispersed phases increases in the near-wall region. On the one hand, this is attributed to the fact that the spectrum of fluctuations of air velocity in the normal direction is characterized by higher frequencies [18], and the particles are less entrained by turbulent eddies of the carrier phase. On the other hand, the averaged velocity of particles in the direction being treated is close to zero in the entire cross-section of the boundary layer. Consequently, possible movements of the dispersed phase in the transverse direction do not result in the emergence of "additional" fluctuations (as was the case with longitudinal fluctuations).

Rogers and Eaton [20] attempted a study of the effect of particles on the characteristics of a turbulent boundary layer developing on a flat plate. Copper particles $70\,\mu\mathrm{m}$ in diameter at a mass concentration $M = 0.2$ were used in experiments. The observed effect of particles on the distributions of averaged and fluctuation velocities of carrier air was insignificant and did not exceed the experimental error. This is attributed to the relatively low concentration of the dispersed phase. In spite of this, the presence of particles made an impact on the spectrum of longitudinal fluctuations of gas velocity by suppressing the low- frequency components (Fig. 5.37).

A heterogeneous boundary layer was experimentally investigated in detail in [26, 29–31]. The boundary layer developed along the side surface of a rod with a hemispherical end placed within a vertical pipe (see Fig. 3.20). The rod was subjected to an upward flow of air. The dispersed phase was provided in the experiments by particles of alumina (Al_2O_3) with a mass average size of

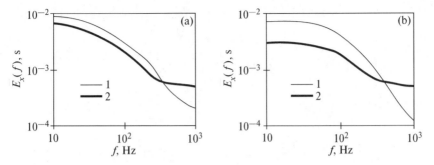

Fig. 5.37. The effect of the presence of copper particles (70 μm) on the energy spectrum of air turbulence in a turbulent boundary layer ($x = 0.85$ m, $U_{x0} = 8$ m s^{-1}, $Re_x = 4.5 \times 10^5$): **(a)** $y^+ = 100$ **(b)** $y^+ = 300$; (1) $M = 0$, (2) $M = 0.2$

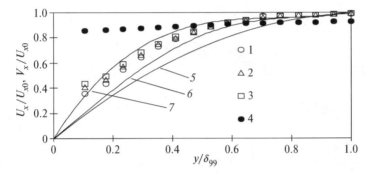

Fig. 5.38. The distributions of averaged velocities in a "pseudolaminar" boundary layer ($Re_x = 2.1 \times 10^4, U_{x0} = 13.3$ m s^{-1}): (1)air ($M = 0$), (2) air ($M = 0.18$), (3) air ($M = 0.26$), (4) Al$_2$O$_3$ particles, (5) theoretical Blasius profile ($\sigma_U = 0$), (6) $\sigma_U = 3.66\%$, (7) $\sigma_U = 7.79\%$

50 μm. The average (over the pipe cross-section) mass flow-rate concentration of particles was varied in the range $\langle M_G \rangle \approx M = 0 - 0.26$. It was assumed that, given such parameters of particles and concentrations, the effect of the dispersed phase on incident air flow must be minimal. The measurements performed in an incident (unperturbed by the rod) flow revealed that the presence of particles does not cause a variation of the profile of averaged velocity of carrier air flow.

Profiles of averaged velocities of "pure" air, of air in the presence of particles, and of solid particles proper were measured in all regions of the boundary layer, namely, in the laminar, transition, and turbulent ones. Examples of obtained distributions of velocities are given in Figs. 5.38–5.40. The data in these figures correspond to the Reynolds number $Re_D = 5.5 \times 10^4$. One can see in all graphs that the velocity of particles in the vicinity of the boundary of the boundary layer is lower than the carrier gas velocity and amounts to approximately 90% of the velocity of external flow. At the same time, the velocity

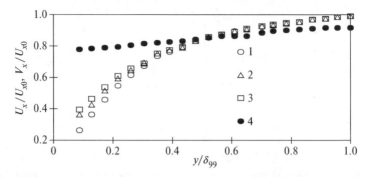

Fig. 5.39. The distributions of averaged velocities in the transition region of a boundary layer ($Re_x = 7 \times 10^4, U_{x0} = 13.3\,\mathrm{m\,s}^{-1}$): (1)air ($M = 0$), (2) air ($M = 0.18$), (3) air ($M = 0.26$), (4) Al$_2O_3$ particles

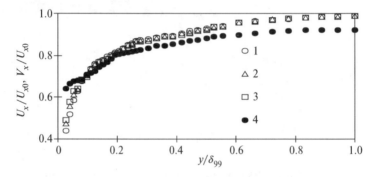

Fig. 5.40. The distributions of averaged velocities in a turbulent boundary layer ($Re_x = 1.54 \times 10^5, U_{x0} = 13.3\,\mathrm{m\,s}^{-1}$): (1)air ($M = 0$), (2) air ($M = 0.18$), (3) air ($M = 0.26$), (4) Al$_2O_3$ particles

of particles in the vicinity of the wall because of their inertia exceeds significantly the velocity of air throughout the investigated region in the boundary layer and, especially so, in the region of laminar flow.

The laminar profile of single-phase flow was much flatter than the classical theoretical Blasius profile ($\sigma_U \to 0$) because of the turbulence of external flow (under conditions of these experiments, $\sigma_{U_x} = (\overline{u'_x})^{1/2}/U_{x0} \approx 6.5\%$). The conventional Blasius profile for a laminar boundary layer of single- phase flow on a flat plate is represented in the coordinates $U_x/U_{x0} = f(\varphi)$, where $\varphi = y\sqrt{U_{x0}/\nu x}$. At $\varphi = 5.0$, the relative velocity $U_x/U_{x0} = 0.99$, i.e., this value of φ defines the boundary layer thickness δ_{99}, namely, $\delta_{99}/x = 5/\sqrt{Re_x}$. When δ_{99} is determined in this conventional manner, the prolate boundary-layer coordinate φ is related to the relative coordinate $\overline{y} = y/\delta_{99}$ by the relation $\varphi = 5(y/\delta_{99}) \equiv 5\overline{y}$. In Fig. 5.38, the Blasius profile is plotted for the experimentally obtained value of $\delta_{99} = 0.85$ mm. The measurement cross-section is spaced at a distance $x = 23.6$ mm (which corresponds to $Re_x = 2.1 \times 10^4$) reckoned from the forward stagnation point along the surface subjected to flow. Given the validity of the foregoing correlations between δ_{99}/x and Re_x, which are characteristic of an unperturbed flow past a flat plate, the calculated value is $\delta_{99} = 0.81\,\text{mm}$; this is only several percent different from the measured value on a hemispherically blunted rod. This fits the literature data of [9] according to which the turbulization of external flow must cause an increase in the thickness of laminar boundary layer.

Also given in Fig. 5.38 is the distribution of velocity in a "pseudolaminar" single-phase boundary layer for the degree of turbulence (intensity of fluctuations) in incident flow of $\sigma_U = 3.66\%$ and 7.79% according to the data of [9] for a similar value of the Reynolds number $Re_x = 2 \cdot 10^4$. Dyban and Epik [9] define the laminar boundary layer in a turbulized flow as "pseudolaminar," because it is characterized by intensive fluctuations of local parameters. In this layer, the predominating influence of molecular viscosity is retained; the equilibrium region of generation and dissipation of turbulence (which is characteristic of a turbulent boundary layer), i.e., the region of logarithmic law of the wall, is not realized in the layer. One can readily observe that the data obtained for a single-phase flow are located between the respective data of [9] for the velocity distribution in a laminar boundary layer and, therefore, agree with these data.

The presence of particles in the flow has a significant effect on the profile of averaged velocity of the carrier phase in the "pseudolaminar" boundary layer (see Fig.5.38). The profile becomes flatter because of the acceleration of air by particles in the vicinity of the wall. This fits the inferences made by Osiptsov [15]. The difference between the velocities of single-phase and heterogeneous flows reaches its maximum precisely in the vicinity of the wall, where the maximal difference is observed between the velocities of the gas and solid phases because of the inertia of particles. The Reynolds number of a particle in this region exceeds significantly the respective characteristic in incident flow and is $Re_p = 15\text{--}25$. The particles make flatter the profile of averaged velocity and cause an increase of the gradient of this velocity on the wall; this leads to an increase in friction in the laminar region of the boundary layer. The flattening of the velocity profile further leads to a decrease in the form factor of profile in this region and brings its value closer to that which is characteristic of a turbulent boundary layer; therefore, it expedites the beginning of alternating laminar-turbulent flow.

Two methods were used to determine the beginning of laminar-turbulent transition, namely, (1) by the minimum in the distribution of averaged velocity when traversing the measurement point at a fixed distance along the rod surface, and (2) by the beginning of an abrupt decrease in the form factor of the profile of averaged velocity.

The determination of the beginning of laminar-turbulent transition by the minimum of averaged velocity is illustrated by Fig. 5.41. The distance from the wall ($y = 0.2\,\text{mm}$), at which the traversing along the surface was performed, was selected experimentally so that the minimum would be defined

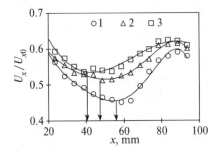

Fig. 5.41. The determination of the coordinate of the beginning of laminar turbulent transition ($U_{x0} = 13.3\,\mathrm{m\,s^{-1}}$): (1) $M = 0$, (2) $M = 0.18$, (3) $M = 0.26$

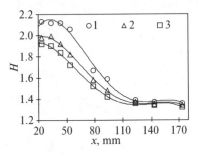

Fig. 5.42. The distributions of the form factor of the profile of averaged velocity ($U_{x0} = 13.3\,\mathrm{m\,s^{-1}}$): (1) $M = 0$, (2) $M = 0.18$, (3) $M = 0.26$

most clearly. The distribution of the form factor H of the velocity profile is given in Fig. 5.42. This characteristic is determined as follows:

$$H = \frac{\delta^*}{\delta^{**}}, \tag{5.14}$$

where the displacement thickness δ^* and the momentum thickness δ^{**} are found for incompressible flow as

$$\delta^* = \int_0^{\delta_{99}} \left(1 - \frac{U_x}{U_{x0}}\right) dy, \delta^{**} = \int_0^{\delta_{99}} \frac{U_x}{U_{x0}} \left(1 - \frac{U_x}{U_{x0}}\right) dy \tag{5.15}$$

One can see in Fig. 5.42 that the value of the form factor of the velocity profile under conditions of single-phase flow is much lower than that for the classical laminar boundary layer ($H = 2.6$, $\sigma_U \to 0$) and is $H \approx 2.10$–2.15. This is due to the higher degree of turbulence in the flow core. As was already mentioned above, the presence of particles in the flow results in a flatter profile of averaged velocity and, thereby, causes a decrease in the value of the form factor to $H \approx 2.0$ at the concentration of the dispersed phase $M = 0.18$ and to $H \approx 1.90$–1.92 at the concentration $M = 0.26$.

In analyzing Figs. 5.41 and 5.42, one can infer that, in the case of single-phase flow, the laminar-turbulent transition begins at $x = 55$–60 mm, which corresponds to the values of the Reynolds number in the range $Re_{x\,\mathrm{cr1}} = 4.88{\times}10^4$–$5.32{\times}10^4$. For a heterogeneous flow, the laminar-turbulent transition begins earlier, namely, at $x = 40$–50 mm ($Re_{x\,\mathrm{cr1}} = 3.55{\times}10^4$–$4.43{\times}10^4$) and $x = 36$–44 mm ($Re_{x\,\mathrm{cr1}} = 3.19{\times}10^4$–$3.90{\times}10^4$) for mass concentrations $M = 0.18$ and $M = 0.26$, respectively. Note that it is in the case of determining the coordinate of the point of beginning of laminar-turbulent transition that the above-described fact of actual absence of the effect of particles on the structure of incident flow assumes great importance. It is well known that, in the case of single-phase flow, a decrease in the intensity of turbulence of incident flow causes a significant delay of the beginning of laminar-turbulent transition. For the performed experiments, the degree of turbulence of "pure" air was $\sigma_{U_x} \approx 6.5\%$ in the flow core; in the presence of particles, this degree decreased to $\sigma_{U_x} \approx 5.6\%$; in accordance with the available literature data, this should not have resulted in a significant increase in the first critical Reynolds number $Re_{x\,\mathrm{cr1}}$. Therefore, the experiments revealed the impact made by the presence of solid particles on the beginning of laminar-turbulent transition.

Figure 5.39 demonstrates the distributions of averaged velocities of "pure" air and of both phases of heterogeneous flow in the transition region of the boundary layer. In spite of the fact that the difference between the velocities of the phases in the wall region decreases, the turbulizing effect of particles on the alternating laminar-turbulent flow is as intensive as in the case of "pseudolaminar" boundary layer. This effect is reflected by the significant flattening of the profile of carrier gas velocity. As to the effect of particles on the termination of laminar-turbulent transition, no such effect was revealed.

Profiles of averaged velocities of "pure" air and of both phases of heterogeneous flow in a turbulent boundary layer are given in Fig. 5.40. The measured distributions of velocities in the turbulent boundary layer are within the relaxation region, i.e., the region where the velocity of particles decreases downstream. For example, in the case of the cross-section shown in Fig. 5.40, the velocity of the dispersed phase in the vicinity of the wall still remains much higher than the velocity of single-phase flow. As a result, the qualitative pattern of distributions of velocities of the gas and solid phases in the turbulent boundary layer remains the same as that in the cases shown in Figs. 5.38 and 5.39.

Measurements were further made of profiles of the degree of turbulence in a turbulent boundary layer for selected cross-sections. Figure 5.43 gives a typical distribution of the intensity of turbulence for the cross-section $x = 173.6$ mm (the respective distributions of averaged velocities are given in Fig. 5.40). One can see that the presence of particles leads to a significant suppression of the intensity of turbulent fluctuations of the carrier phase, which reaches its maximum in the immediate vicinity of the wall.

It is known that the main mechanism of dissipation of turbulence by particles is their involvement in the fluctuation motion by turbulent eddies. The

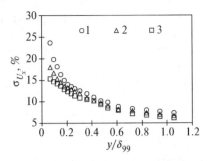

Fig. 5.43. The distributions of the intensity of turbulence of air in a turbulent boundary layer ($Re_x = 1.54 \times 10^5$, $U_{x0} = 13.3\,\mathrm{m\,s^{-1}}$): (1) $M = 0$, (2) $M = 0.18$, (3) $M = 0.26$

experimentally observed fact of maximal suppression of turbulent fluctuations in the vicinity of the wall merits attention, because it is not obvious how turbulent eddies with short characteristic lifetimes (which are observed in the wall region) manage to significantly affect the velocity of rather inertial particles. We will try to explain the experimentally observed high dissipation of turbulence using the results of analysis of the equation of transfer of kinetic energy of carrier air, which will be written in the following simplified form:

$$\frac{Dk}{D\tau} = P - \varepsilon - \varepsilon_\mathrm{p}, \tag{5.16}$$

where $P = \nu_\mathrm{t}(\partial U_x/\partial y)^2$ is the term responsible for the generation of turbulence; ε is the term responsible for the dissipation of turbulence (in a first approximation, this term may be taken to be equal to the respective term for single-phase flow); and ε_p is the term defining the additional dissipation of turbulence due to the presence of particles. The expression for ε_p is written as

$$\varepsilon_\mathrm{p} = \frac{1}{\tau_\mathrm{p}} \sum_i [M(\overline{u_i'u_i'} - \overline{u_i'v_i'}) + (U_i - V_i)\overline{m'u_i'} + (\overline{m'u_i'u_i'} - \overline{m'u_i'v_i'})]. \tag{5.17}$$

In the case of flow with Stokesian particles, the second term in the right-hand part of the expression for ε_p is small compared to the first term. The third term containing triple correlations is also ignored usually. As applied to the conditions of the given investigations, the presence of two main factors is assumed which cause a higher dissipation of turbulence compared to the case of flow with Stokesian particles. These factors include (1) the presence of significant difference between the averaged velocities of the phases causes an intensive exchange of momentum, flattening of the profile of averaged velocity of the carrier phase in the near-wall region ($\bar{y} = 0.03 - 0.1$), a decrease in the velocity gradient in this region, and a decrease in the generation of turbulence P; and (2) the presence of dynamic slip in averaged motion in the longitudinal direction causes an increase in additional dissipation ε_p due to the increase in the first term on the right-hand side of (5.17).

5.4 The Body Drag in Particle-Laden Flows

Investigations of aerodynamic drag of bodies moving in a heterogeneous flow include experimental studies [1–3].

The drag of flat aluminum wedges with apex angles $\alpha = 10\text{–}180\,^\circ$ in a flow of air ($U_{x0} = 200\,\mathrm{m\,s^{-1}}$) with particles of alumina ($\mathrm{Al_2O_3}$, $\rho_\mathrm{p} = 3{,}900\,\mathrm{kg\,m^{-3}}$) of mass average sizes $d_\mathrm{p} = 16\text{–}88\,\mu\mathrm{m}$ was studied by Balanin and Lashkov [2]. The cross-section average mass flow-rate concentration of particles was measured in the range $\langle M_\mathrm{G}\rangle = 0\text{–}0.3$. A strain-gauge balance was used to measure the drag force in a two-phase flow. It was assumed that the total force F_Σ acting on the model in a heterogeneous flow consisted of two independent terms, namely, the gas phase impact force F_0 and the particle impact force F_p,

$$F_\Sigma = F_0 + F_\mathrm{p} \tag{5.18}$$

If the drag coefficient C_{xp} due to the impact of particles is introduced, (5.18) gives

$$\frac{F_\Sigma}{F_0} = 1 + \frac{C_{xp}\rho_\mathrm{p}\langle\varPhi\rangle\langle V_x^2\rangle}{C_{x0}\rho\langle U_x^2\rangle}, \tag{5.19}$$

where C_{x0} is the coefficient of drag due to the effect of "pure" air; $\langle U_x\rangle$ and $\langle V_x\rangle$ denote the model cross-section average velocities of air and particles, respectively; and $\langle\varPhi\rangle$ is the volume concentration of particles.

The relative drag force F_Σ/F_0 as a function of the concentration of particles of different sizes for a wedge with an angle of 120° is given in Fig. 5.44.

Equation (5.19) yields the following relation:

$$\frac{C_{xp}}{C_{x0}} = \left(\frac{F_\Sigma}{F_0} - 1\right)\frac{\rho\langle U_x^2\rangle}{\rho_\mathrm{p}\langle\varPhi\rangle\langle V_x^2\rangle}. \tag{5.20}$$

Expression (5.20) was used to determine the drag coefficient C_{xp} which characterizes the force due to particle impacts in terms of the experimentally measured values of the forces F_0 and F_Σ.

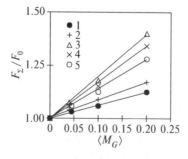

Fig. 5.44. The relative drag force as a function of the concentration of particles for a wedge with an angle of 120°: (1) $d_\mathrm{p} = 16\,\mu\mathrm{m}$, (2) $23\,\mu\mathrm{m}$, (3) $32\,\mu\mathrm{m}$, (4) $44\,\mu\mathrm{m}$, (5) $88\,\mu\mathrm{m}$

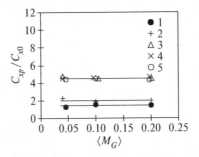

Fig. 5.45. The relative drag coefficient as a function of the concentration of particles for a wedge with an angle of 120°: (1) $d_p = 16 \, \mu$m, (2) 23 μm, (3) 32 μm, (4) 44 μm, (5) 88 μm

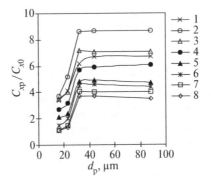

Fig. 5.46. The relative drag coefficient as a function of the particle size for different angles of wedges: (1) 10°, (2) 20°, (3) 40°, (4) 60°, (5) 90°, (6) 120°, (7) 150°, (8)180°

Figure 5.45 gives the relative drag coefficient C_{xp}/C_{x0} as a function of the concentration of particles. One can see in this figure that this characteristic is independent of the concentration of particles in a heterogeneous flow.

The relative drag coefficient as a function of the particle size for different angles of wedges is given in Fig. 5.46. Two regions may be identified on all of the distribution curves, the boundary between which falls on the particles of $d_{pcr} \approx 30 \, \mu$m. When wedges are subjected to heterogeneous flow with particles of size $d_p < d_{pcr}$, a strong dependence of the drag coefficient on the size of the dispersed phase is observed. For a flow with particles of $d_p > d_{pcr}$, the transfer of momentum from the dispersed phase to the model (aerodynamic drag) ceases to depend on the particle sizes for all angles of wedges.

The effect of the wedge angle on its resistance (Fig. 5.47) turns out to be rather unexpected (as was observed in [2]). These data lead one to infer that the maximal effect of the solid phase is attained at an angle $\alpha = 20°$ for any sizes of particles, and the relative drag of blunt bodies in a heterogeneous flow is lower than that of more pointed bodies.

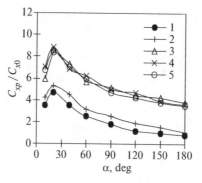

Fig. 5.47. The relative drag coefficient as a function of the wedge angle: (1) $d_\mathrm{p} = 16\,\mu\mathrm{m}$, (2) $23\,\mu\mathrm{m}$, (3) $32\,\mu\mathrm{m}$, (4) $44\,\mu\mathrm{m}$, (5) $88\,\mu\mathrm{m}$

One can see in the latter two figures that the aerodynamic drag of a body in a two-phase flow, which was caused by the effect of particles alone (for certain particle sizes and wedge angles), exceeded the drag in "pure" air by a factor of 8. Therefore, the data given in Figs. 5.46 and 5.47 are indicative of the existence of a critical size and a critical angle in the case of which the process of flow past a body changes qualitatively.

The rise of aerodynamic drag of bodies in heterogeneous flows is defined primarily by the process of particle/surface interaction. The intensity of this process is directly dependent on the sedimentation factor which (as was demonstrated above) is largely defined by the inertia of particles (by the Stokes number). We will try to analyze the results described above by estimating the Stokes number for these experiments,

$$Stk_\mathrm{f} = \frac{\tau_\mathrm{p}}{T_\mathrm{f}}, \qquad (5.21)$$

where $\tau_\mathrm{p} = \tau_\mathrm{p0}/C = \rho_\mathrm{p} d_\mathrm{p}^2/18\mu C$, $C = 1 + Re_\mathrm{p}^{2/3}/6$. The Reynolds number of a particle will be found using the data given in [2] on the dynamic slip of particles in an unperturbed flow, i.e., $\langle V_x \rangle / \langle U_x \rangle = 0.75 - 0.58$ for particles of $d_\mathrm{p} = 16$–$88\,\mu\mathrm{m}$.

The characteristic time of carrier gas will be estimated as

$$T_\mathrm{f} = \frac{L}{U_{x0}}, \qquad (5.22)$$

where $L = h/\sin\alpha/2$ ($h = 5\,\mathrm{mm}$ is a half-height of the cone base).

Figure 5.48 gives the distribution of the Stokes number calculated by relations (5.21) and (5.22) as a function of the particle size and of the wedge angle. The number of particles interacting with the body surface increases with the Stokes number. This leads to an increase in the aerodynamic drag. The effectiveness of this mechanism terminates when the sedimentation factor ceases to increase. We will assume for estimation that $\eta \approx 1$ at $Stk_\mathrm{f} \geq 10$.

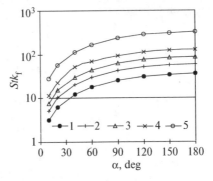

Fig. 5.48. The Stokes number of particles of different sizes as a function of the wedge angle: (1) $d_p = 16\,\mu\mathrm{m}$, (2) $23\,\mu\mathrm{m}$, (3) $32\,\mu\mathrm{m}$, (4) $44\,\mu\mathrm{m}$, (5) $88\,\mu\mathrm{m}$

Then, the data given in Fig. 5.48 lead one to infer that, in the case of particles of $d_p \geq 32\,\mu\mathrm{m}$ and wedge angles $\alpha \geq 20°$, the sedimentation factor is $\eta \approx 1$. Therefore, when the above-identified values of the size of particles and of the angle of wedges are attained, the increase in the aerodynamic drag due to collisions between the dispersed phase and the surface of bodies subjected to flow must cease.

In addition to direct interaction between the particles and the wall, the drag coefficient of a body may be affected by processes such as the inverse effect of particles on the parameters of carrier gas and on collisions between particles. Due to their inertia, the particles incident on a body exhibit a higher velocity than the decelerating gas. As a result, the particles accelerate the gas to cause an increase in the body drag. On the contrary, the particles reflected from a body move toward the gas; this is a factor causing a decrease in the gas velocity and in the coefficient of aerodynamic drag of the body. Note that the particles reflected from a body apparently make a stronger impact (compared to the incident particles) on the parameters of gas flow, because these particles move under conditions in which the dynamic slip between the phases is several times higher. The number of reflected particles and their effect on the carrier phase increases abruptly with increasing size (inertia) of the dispersed phase and increasing wedge angle; this leads to a decrease in the aerodynamic drag. This inference is supported by the results of experiments in rebound of inertial particles given in Sect. 5.2. Apparently, the experimentally observed decrease in the relative resistance of blunt bodies compared to that of pointed ones is associated with the mechanism of inverse effect of particles on gas described above.

Conclusions

The characteristics of turbulent flows of air in the presence of solid particles in pipes (channels) and under conditions of flow past bodies have been treated. The basic results given in the book include:

1. The suggested classification of turbulent heterogeneous flows by the volume concentration and inertia (Stokes number) of particles in averaged, large-scale, and small-scale fluctuation motion.
2. The solution of a large complex of metrological problems associated with the diagnostics of the structure of particle-laden turbulent flows of gas.
3. Detailed analysis of the characteristics of motion of particles and of their inverse effect on the parameters of carrier gas under conditions of flow in vertical pipes and of flow past bodies.
4. The description (based on unified methodology within a single mathematical model) of the processes of additional dissipation and additional generation of turbulence in heterogeneous flows with relatively low- inertia and large particles, respectively.

Note at the same time that many of the important problems and aspects of the theory of turbulent flows of gas with solid particles have been hardly dealt with in the book. These are the characteristic features of highly dust-laden flows, the regularities of formation of profiles of concentration of the dispersed phase, heterogeneous flows in horizontal channels, the characteristic features of high-velocity flows with solid particles, the sedimentation of particles on channel walls, and many others.

One can use the suggested classification of turbulent heterogeneous flows to estimate in advance (prior to investigations) the presence and intensity of determining interphase interactions and exchange processes. The thus developed classification may be recommended for use in theoretical and experimental investigations of multiphase flows of various types.

The investigations referred to in the book are not characterized by a clearly defined application pattern; therefore, the results of these investigations will find application in the most diverse spheres of human activities. We will take a brief look at possible spheres of practical utilization of these results.

The developed procedure of measurements in heterogeneous flows opens up extensive possibilities for improvements in the diagnostics of multiphase flows. The diagnostics of heterogeneous flows pursue two objectives, namely, those of (1) determining the characteristics of flow for the purpose of maintaining optimal process conditions and (2) obtaining data to be employed in the calculations of concrete processes.

Examples of devices in which the thus developed diagnostic procedures may be employed to advantage include devices for sand- and shot-blasting of various surfaces, pneumatic conveyers of loose materials, classifiers of polydisperse materials by the particle size, various dust collectors, devices for thermal preparation of coal in schemes for energy-technology utilization of fuel, combustors of heat engines, devices for heat treatment of loose materials, heat exchangers with two-phase working media, and so on and so forth.

Such procedures may be further recommended for use in performing measurements in flows of gas suspensions resulting from natural processes (fogs, sandstorms, forest fires, volcanic eruptions, etc.) and from human activities (production of dust and harmful exhausts by moving vehicles and atmospheric pollution by industrial emissions).

Priority problems associated with the diagnostics of turbulent heterogeneous flows today include:

1. The investigation of the characteristics of highly dust-laden flows which involve collisional interaction of particles.
2. The measurements of the fields of correlation of fluctuations of velocities of the carrier and dispersed phases, correlations of fluctuations of concentration and velocity of particles, and so on.
3. The investigation of the impact made by particles on the fine structure of turbulence of carrier flow, in particular, on the spectrum of turbulent fluctuations of velocity and on the microscales of turbulence.
4. The investigation of the statistical parameters of flows with substantially polydisperse particles, as well as with bidisperse, hollow, porous particles, etc.
5. The development of methods of local measurement of the temperature of continuum flow, as well as of fluctuations of this temperature in the presence of particles.

In view of the rate of development of the methods of experimental investigation of heterogeneous flows, as well as of ever growing interest of numerous research teams in studying such flows, one can hope that the problems listed above will be solved in the nearest future.

The investigation results make it possible to predict the effect of particles on the turbulent energy of carrier flow. The understanding of the physics of interaction between particles and surrounding gas further enables one to control the integral characteristics of flow, such as friction and heat transfer.

The control of the properties of continuum flows in flow trains of power plants by introducing particles of certain physical properties into the flow at certain concentration of particles may be very effective. However, one must bear in mind that the presence of particles in flows almost always entails the possibility of sedimentation of particles on the walls, erosion, and other negative effects.

The investigation results described in the book make it possible to raise the intensity of processes occurring in heat exchangers, which utilize gas suspensions, by optimal selection of the basic structural (pipe diameter, etc.) and process (gas velocity) parameters, as well as of the properties (thermal and dynamic inertia) of the employed particles.

Examples of practical uses of "gas–solid particles" heterogeneous flows may be further provided by pulverized-coal ducts and pneumatic conveyers of loose materials; in the nearest future, such devices may possibly present an alternative to automobile and railroad transport. The consumption of energy for the pipeline conveyance of various materials are directly associated with the pressure loss due to a number of reasons such as the roughness of pipe walls, the conveyer length, the pipe diameter, the density, viscosity, and velocity of gas, the type of material being conveyed, the density, size, and velocity of particles, and so on. Investigations revealed that, in addition to the above-identified parameters taken into account in designing pneumatic-conveyance systems, one must further take into account the Stokes number of particles of the material being conveyed in large-scale fluctuation motion. The correct choice of this characteristic of heterogeneous flow will make it possible to significantly lower the level of turbulent fluctuations of velocity of the carrier gas phase and, consequently, reduce the hydraulic resistance and loss of pressure due to conveyance.

The process of dissolution of solid particles is a typical mass-exchange process extensively employed in the chemical industry. The inclusion of the modification of turbulence by particles will make it possible to optimize column-type units for dissolving particles in a fluidized bed; this will raise significantly the efficiency of this process.

The foregoing may be extended to various devices capable of heat treatment of loose materials (heating, drying, cooling, sintering, etc.) and employed in the food, medical, and other industries.

Note that the model of additional generation and dissipation of turbulent energy of gas flow by particles, described in this book, may be used in calculating and designing various turbulence stimulators and damping grids extensively employed in power generation and in aviation and space engineering.

I hope that this monograph will generate interest among students, postgraduates, and researchers involved in investigations of hydrodynamics and heat transfer in solid particle-carrying heterogeneous flows and will give impetus to further development of the theory of multiphase flows.

References

Preface

1. Abramovich, G.N., Girshovich, T.A., Krasheninnikov, S.Yu, et al.: Teoriya turbulentnykh strui (The Theory of Turbulent Jets). Nauka, Moscow (1984)
2. Babukha, G.L., Rabinovich, M.I.: Mekhanika i teploobmen potokov polidispersnoi gazovzvesi (The Mechanics and Heat Transfer of Flows of Polydisperse Gas Suspension). Naukova Dumka, Kiev (1968)
3. Babukha, G.L., Shraiber, A.A.: Vzaimodeistvie chastits polidispersnogo materiala v dvukhfaznykh potokakh (The Interaction of Particles of Polydisperse Material in Two-Phase Flows). Naukova Dumka, Kiev (1972)
4. Boothroyd, R.G.: Flowing Gas-Solids Suspensions. Chapman & Hall, London (1971)
5. Clift, R., Grace, J.R., Weber, M.E.: Bubbles, Drops, and Particles. Academic, New York (1978)
6. Crowe, C., Sommerfeld, M., Tsuji, Y.: Multiphase Flows with Droplets and Particles, CRC, Boca Raton (1998)
7. Deich, M.E., Filippov, G.A.: Gazodinamika dvukhfaznykh sred (The Gas Dynamics of Two-Phase Media). Energoizdat, Moscow (1981)
8. Dyunin, A.K., Borshchevskii, Yu.T., Yakovlev, N.A.: Osnovy mekhaniki mnogokomponentnykh potokov (Fundamentals of the Mechanics of Multicomponent Flows) SO AN SSSR (Siberian Div., USSR Acad. Sci.). Novosibirsk (1965)
9. Friedlander, S.K.: Smoke, Dust and Haze: Fundamentals of Aerosol Behavior. Wiley, New York (1977)
10. Fuks, N.A.: Mekhanika aerozolei (Aerosol Mechanics) Izd. AN SSSR (USSR Acad. Sci.). Moscow (1955)
11. Fuks, N.A.: Uspekhi mekhaniki aerozolei (Progress in Aerosol Mechanics) Izd. AN SSSR (USSR Acad. Sci.). Moscow (1961)
12. Gorbis, Z.R.: Teploobmen dispersnykh skvoznykh potokov (Heat Transfer in Disperse through Flows). Energiya, Moscow-Leningrad (1964)
13. Gorbis, Z.R.: Teploobmen i gidromekhanika dispersnykh skvoznykh potokov (Heat Transfer and Fluid Mechanics in Disperse through Flows). Energiya, Moscow (1970)

14. Gorbis, Z.R., Kalender'yan, V.A.: Teploobmenniki s protochnymi dispersnymi teplonositelyami (Heat Exchangers with Flow-Through Heat-Transfer Agents). Energiya, Moscow (1975)
15. Grishin, A.M., Fomin, V.M.: Sopryazhennye i nestatsionarnye zadachi mekhaniki reagiruyushchikh sred (Conjugate and Unsteady-State Problems in the Mechanics of Reacting Media). Nauka, Novosibirsk (1984)
16. Hetsroni, G. (ed.): Handbook of Multiphase Systems. McGraw-Hill, New York (1982)
17. Mednikov, E.P.: Turbulentnyi perenos i osazhdenie aerozolei (Turbulent Transport and Sedimentation of Aerosols). Nauka, Moscow (1981)
18. Nigmatulin, R.I.: Osnovy mekhaniki geterogennykh sred (The Fundamentals of Mechanics of Heterogeneous Media). Nauka, Moscow (1978)
19. Nigmatulin, R.I.: Dinamika mnogofaznykh sred (The Dynamics of Multiphase Media) (Parts 1 and 2). Nauka, Moscow (1987)
20. Saltanov, G.A.: Sverkhzvukovye dvukhfaznye techeniya (Supersonic Two-Phase Flows). Vysheishaya Shkola, Minsk (1972)
21. Saltanov, G.A.: Neravnovesnye i nestatsionarnye protsessy v gazodinamike odnofaznykh i dvukhfaznykh sred (Nonequilibrium and Unsteady-State Processes in the Gas Dynamics of Single-Phase and Two-Phase Media). Nauka, Moscow (1979)
22. Shraiber, A.A., Gavin, L.B., Naumov, V.A., Yatsenko, V.P.: Turbulentnye techeniya gazovzvesi (Turbulent Flows of Gas Suspension). Naukova Dumka, Kiev (1987)
23. Shraiber, A.A., Milyutin, V.N., Yatsenko, V.P.: Gidromekhanika dvukhkomponentnykh potokov s tverdym polidispersnym veshchestvom (The Fluid Mechanics of Two-Component Flows with Solid Polydisperse Matter). Naukova Dumka, Kiev (1980)
24. Soo, S.L.: Fluid Dynamics of Multiphase Systems. Blaisdell, Waltham (1967)
25. Soo, S.L.: Particulates and Continuum. Multiphase Fluid Dynamics. Hemisphere, New York (1989)
26. Sternin, L.E.: Osnovy gazodinamiki dvukhfaznykh techenii v soplakh (The Fundamentals of Gas Dynamics of Two-Phase Flows in Nozzles). Mashinostroenie, Moscow (1974)
27. Sternin, L.E., Maslov, B.N., Shraiber, A.A., Podvysotskii, A.M.: Dvukhfaznye mono- i polidispersnye techeniya gaza s chastistsami (Two-Phase Mono- and Polydisperse Flows of Gas with Particles). Mashinostroenie, Moscow (1980)
28. Sternin, L.E., Shraiber, A.A.: Mnogofaznye techeniya gaza s chastistsami (Multiphase Flows of Gas with Particles). Mashinostroenie, Moscow (1994)
29. Sukomel, A.S., Tsvetkov, F.F., Kerimov, R.V.: Teploobmen i gidravlicheskoe soprotivlenie pri dvizhenii gazovzvesi v trubakh (Heat Transfer and Hydraulic Resistance under Conditions of Gas Suspension Motion in Pipes). Energiya, Moscow (1977)
30. Volkov, E.P., Zaichik, L.I., Pershukov, V.A.: Modelirovanie goreniya tverdogo topliva (Simulation of Solid Fuel Combustion). Nauka, Moscow (1994)
31. Voloshchuk, V.M.: Vvedenie v gidrodinamiku grubodispersnykh aerozolei (Introduction to Fluid Dynamics of Coarse Aerosols). Gidrometeoizdat, Leningrad (1971)
32. Voloshchuk, V.M., Sedunov, Yu.S.: Protsessy koagulyatsii v dispersnykh sistemakh (Processes of Coagulation in Disperse Systems). Gidrometeoizdat, Leningrad (1975)

33. Wallis, G.: One-Dimensional Two-Phase Flow. McGraw-Hill, New York (1969)
34. Yanenko, N.N., Soloukhin, R.I., Papyrin, A.N., Fomin, V.M.: Sverkhzvukovye dvukhfaznye techeniya v usloviyakh skorostnoi neravnovesnosti chastits (Supersonic Two-Phase Flows under Conditions of Velocity Disequilibrium of Particles). Nauka, Novosibirsk (1980)

Chapter 1

1. Baranovskii, S.I.: Characteristic features of high-velocity two-phase gas–liquid flows. Turbulentnye dvukhfaznye techeniya i tekhnika eksperimenta (Turbulent Two-Phase Flows and Experimental Techniques), Tallinn, p. 60 (1985)
2. Bradshaw, P.: An Introduction to Turbulence and Its Measurement. Pergamon, London (1971)
3. Bradshaw, P. (ed.): Turbulence. Springer, Berlin Heidelberg New York (1971)
4. Comte-Bellot, G., Ecoulement turbulent entre deux parois paralleles, Publ. Scientifique et Technique du Ministere de l'Armee de l'Air **419** (1965)
5. Crowe, C.T.: Review – numerical models for dilute gas-particles flows. Trans. ASME J. Fluids Eng. **104**(3), 297 (1982)
6. Davydov, B.I.: Statistical dynamics of turbulent incompressible fluid. Dokl. Akad. Nauk SSSR **136**(1), 47 (1961)
7. Dyban, E.P., Epik, E.Ya.: Teplomassoobmen i gidrodinamika turbulizovannykh potokov (Heat and Mass Transfer of Turbulized Flows). Naukova Dumka, Kiev (1985)
8. Elghobashi, S.: Particle-Laden turbulent flows: Direct simulation and closure models. Appl. Sci. Res. **48**, 301 (1991)
9. Frost, W., Moulden, T.H.: Handbook of Turbulence. Plenum, New York (1977)
10. Gore, R.A., Crowe, C.T.: Effect of particle size on modulating turbulent intensity. Int. J. Multiphase Flow **15**(2), 279 (1989)
11. Gore, R.A., Crowe, C.T.: Modulation of turbulence by a dispersed phase. Trans. ASME J. Fluids Eng. **113**(2), 304 (1991)
12. Hinze, J.O.: Turbulence: An Introduction to Its Mechanisms and Theory. McGraw-Hill, New York (1959)
13. Kolmogorov, A.N.: Equations of turbulent motion of incompressible fluid. Izv. Akad. Nauk SSSR Ser. Fiz. **6**(1/2), 56 (1942)
14. Kolovandin, B.A.: Correlation Simulation of Transfer Processes in Shear Turbulent Flows. Preprint of Lykov Inst. of Heat and Mass Transfer, Natl Acad Sci Belarus – ITMO, Minsk (1982)
15. Kuznetsov, V.R., Sabel'nikov, V.A.: Turbulentnost' i gorenie (Turbulence and Combustion). Nauka-Fizmatlit, Moscow (1986)
16. Lushchik, V.G., Yakubenko, A.E.: Comparison of models of turbulence for the calculation of wall boundary layer. Izv. Ross. Akad. Nauk Mekh. Zhidk. Gaza **1**, 44
17. Monin, A.S., Yaglom, A.M.: Statisticheskaya gidrodinamika (Statistical Fluid Dynamics). Nauka, Moscow (1965) (Part 1), (1967) (Part 2)
18. Owen, P.R.: Pneumatic transport. J. Fluid Mech. **39**(Pt 2), 407 (1969)
19. Petukhov, B.S., Polyakov, A.F.: Teploobmen pri smeshannoi turbulentnoi konvektsii (Heat Transfer under Conditions of Mixed Turbulent Convection). Nauka, Moscow (1986)

20. Schlichting, H.: Boundary Layer Theory. McGraw-Hill, New York (1968)
21. Varaksin, A.Y., Kurosaki, Y., Satoh, I.: Review: Turbulence modification in gas–solid two-phase wall-bounded flows. Therm. Sci. Eng. **3**(2), 1 (1995)
22. Zaichik, L.I., Pershukov, V.A.: Problems in simulation of gas-dispersion flows with combustion or phase transitions: A review. Izv. Ross. Akad. Nauk Mekh. Zhidk. Gaza **5**, 3 (1996)

Chapter 2

1. Abramovich, G.N.: The impact made by an impurity of solid particles or droplets on the structure of a turbulent gas jet. Dokl. Akad. Nauk SSSR **190**(5), 1052 (1970)
2. Abramovich, G.N., Girshovich, T.A.: The diffusion of heavy particles in turbulent flows. Dokl. Akad. Nauk SSSR **212**(3), 573 (1973)
3. Abramovich, G.N., Girshovich, T.A.: The effect of the size of particles or droplets on the diffusion of impurity in a turbulent jet. Izv. Akad. Nauk SSSR Mekh. Zhidk. Gaza **4**, 18 (1975)
4. Abramovich, G.N., Bazhanov, V.I., Girshovich, T.A.: A turbulent jet with heavy impurities. Izv. Akad. Nauk SSSR Mekh. Zhidk. Gaza **6**, 41 (1972)
5. Abramovich, G.N., Girshovich, T.A., Krasheninnikov, S.Yu, et al.: Teoriya turbulentnykh strui (The Theory of Turbulent Jets). Nauka, Moscow (1984)
6. Ahmed, A.M., Elghobashi, S.: On the mechanisms of modifying the structure of turbulent homogeneous shear flows by dispersed particles. Phys. Fluid. **12**, 2906 (2000)
7. Berlemont, A., Grancher, M.-S., Gousbet, G.: On the lagrangian simulation of turbulence influence on droplet evaporation. Int. J. Heat Mass Transfer **34**(1), 2805 (1991)
8. Boivin, M., Simonin, O., Squires, K.D.: Direct numerical simulation of turbulence modulation by particles in isotropic turbulence. J. Fluid Mech. **375**, 235 (1998)
9. Boivin, M., Simonin, O., Squires, K.D.: On the prediction of gas-solid flows with two-way coupling using large eddy simulation. Phys. Fluid. **12**, 2080 (2000)
10. Bradshaw, P., Cebeci, T., Fernholz, G.-G, et al.: Turbulentnost' (Turbulence). Mashinostroenie, Moscow (1980) (translated into Russian)
11. Chen, C.P., Wood, P.E.: A turbulence closure model for dilute gas-particle flows. Can. J. Chem. Eng. **63**(3), 349 (1985)
12. Derevich, I.V., Zaichik, L.I.: Sedimentation of particles from turbulent flow. Izv. Akad. Nauk SSSR Mekh. Zhidk. Gaza **5**, 96 (1988)
13. Derevich, I.V., Zaichik, L.I.: Equation for the probability density of the velocity and temperature of particles in a turbulent flow simulated by gaussian random field. Prikl. Mat. Mekh. **54**(5), 767 (1990)
14. Derevich, I.V., Yeroshenko, V.M., Zaichik, L.I.: Hydrodynamics and heat transfer of turbulent gas suspension flows in tubes. 1. Hydrodynamics. Int. J. Heat Mass Transfer **32**(1), 2329 (1989)
15. Derevich, I.V., Yeroshenko, V.M., Zaichik, L.I.: Hydrodynamics and heat transfer of turbulent gas suspension flows in tubes. 2. Heat transfer. Int. J. Heat Mass Transfer **32**(1), 2341 (1989)

16. Deutsch, E., Simonin, O.: Large eddy simulation applied to the motion of particles in stationary homogeneous fluid turbulence, in turbulence Modification in Multiphase Flow. ASME **110**, 35 (1991)
17. Eaton, J.K., Fessler, J.R.: Preferential concentration of particles by turbulence. Int. J. Multiphase Flow **20**, 169 (1994)
18. Elghobashi, S.E., Abou-Arab, T.W.: A two-equation turbulence model for two-phase flows. Phys. Fluid. **26**(4), 931 (1983)
19. Fuks, N.A.: Mekhanika aerozolei (Aerosol Mechanics). Izd. AN SSSR (USSR Acad. Sci.), Moscow (1955)
20. Fessler, J.R., Eaton, J.K.: Particle response in a planar sudden expansion flow. Exp. Therm. Fluid Sci. **15**, 413 (1997)
21. Fessler, J.R., Eaton, J.K.: Turbulence modification by particles in a backward-facing step flow. J. Fluid Mech. **394**, 97 (1999)
22. Gavin, L.B., Shraiber, A.A.: Particle-laden turbulent flows of gas. Itogi Nauki Tekh. Ser. Mekh. Zhidk. Gaza **25**, 90 (1991)
23. Gavin, L.B., Naumov, V.A., Shor, V.V.: Numerical investigation of a gas jet with heavy particles using the two-parameter model of turbulence. Prikl. Mekh. Tekh. Fiz. **1**, 62 (1984)
24. Girshovich, T.A., Leonov, V.A.: The effect of disequilibrium of flow on the fluctuation characteristics of a two-phase jet. In: Turbulentnye dvukhfaznye techeniya (Turbulent Two-Phase Flows), Tallinn, Part 1, p. 21 (1982)
25. Jones, W.P.: Prediction Methods for Turbulent Flows. In: Kollman, W. (ed.). Hemisphere, New York (1980)
26. Kondrat'ev, L.V.: Simulation of two-phase turbulent flow in a stabilized region of a pipe. In: Turbulentnye dvukhfaznye techeniya i tekhnika eksperimenta (Turbulent Two-Phase Flows and Experimental Techniques), Tallinn, Part 2, p. 144 (1985)
27. Laats, M.K.: Some objectives and problems associated with the calculation of a jet with heavy particles. In: Turbulentnye dvukhfaznye techeniya (Turbulent Two-Phase Flows), Tallinn, Part 1, p. 49 (1982)
28. Lain, S., Kohnen, G.: Comparison between eulerian and lagrangian strategies for the dispersed phase in nonuniform particle-laden flows. In: Turbulence and Shear Flow Phenomena – 1: First International Symposium, Santa Barbara, p. 277 (1999)
29. Lepeshinskii, I.A., Sovetov, V.A., Chabanov, V.A.: A model of turbulent interaction of the phases of a multiphase multicomponent nonisothermal nonequilibrium jet. In: Turbulentnye dvukhfaznye techeniya i tekhnika eksperimenta (Turbulent Two-Phase Flows and Experimental Techniques), Tallinn, Part 2, p. 42. (1985)
30. Lesieur, M., Metais, O.: New trends in large-eddy simulations of turbulence. Ann. Rev. Fluid Mech. **28**, 45 (1996)
31. Mednikov, E.P.: Turbulentnyi perenos i osazhdenie aerozolei (Turbulent Transport and Sedimentation of Aerosols). Nauka, Moscow (1981)
32. Melville, W.K., Bray, K.N.C.: A model of the two-phase turbulent jet. Int. J. Heat Mass Transfer **22**, 647 (1979)
33. Mostafa, A.A., Mongia, H.C.: On the interaction of particles and turbulent fluid flow. Int. J. Heat Mass Transfer **31**(1), 2063 (1988)
34. Nigmatulin, R.I.: Osnovy mekhaniki geterogennykh sred (The Fundamentals of Mechanics of Heterogeneous Media). Nauka, Moscow (1978)

35. Reeks, M.W.: On a kinetic equation for the transport of particles in turbulent flows. Phys. Fluid. A **3**(3), 446 (1991)
36. Rizk, M.A., Elghobashi, S.E.: A two-equation turbulence model for dispersed dilute confined two-phase flows. Int. J. Multiphase Flow. **15**(1), 119 (1989)
37. Rubinow, S.I., Keller, J.B.: The transverse force on a spinning sphere moving in a viscous fluid. J. Fluid Mech. **11**, 447 (1961)
38. Saffman, P.G.: The lift on a small sphere in a slow shear flow. J. Fluid Mech. **22**, 385 (1965)
39. Shraiber, A.A., Gavin, L.B., Naumov, V.A., Yatsenko, V.P.: Turbulentnye techeniya gazovzvesi (Turbulent Flows of Gas Suspension). Naukova Dumka, Kiev (1987)
40. Sommerfeld, M.: Numerical simulation of the particle dispersion in turbulent flow: the importance of particle lift forces and different particle/wall collision models, in numerical methods for multiphase flows. ASME **91**, 11 (1990)
41. Sommerfeld, M.: Modelling of particle-wall collisions in confined gas-particle flows. Int. J. Multiphase Flow **18**(6), 905 (1992)
42. Sommerfeld, M., Qiu, H.-H.: Characterization of particle-laden, confined swirling flows by phase-doppler anemometry and numerical calculation. Int. J. Multiphase Flow **19**(6), 1093 (1993)
43. Soo, S.L., Ihrig, H.K., El Kouh, A.F.: Experimental determination of statistical properties of two-phase turbulent motion. Trans. ASME J. Basic Eng. **82**(3), 609 (1960)
44. Squires, K.D., Eaton, J.K.: Particle response and turbulence modification in isotropic turbulence. Phys. Fluid. **2**, 1191 (1990)
45. Squires, K.D., Simonin, O.: Application of DNS and LES to dispersed two-phase turbulent flows. In: Proceedings of the 10th Workshop on Two-Phase Flow Predictions, Merseburg, p. 152 (2002)
46. Sternin, L.E., Maslov, B.N., Shraiber, A.A., Podvysotskii, A.M.: Dvukhfaznye mono- i polidispersnye techeniya gaza s chastistsami (Two-Phase Mono- and Polydisperse Flows of Gas with Particles). Mashinostroenie, Moscow (1980)
47. Sundaram, S., Collins, L.R.: A numerical study of the modulation of isotropic turbulence by suspended particles. J. Fluid Mech. **379**, 105 (1999)
48. Varaksin, A.Yu.: To question about fluctuated velocity and temperature of the non- stokesian particles moving in the turbulent flows. In: Heat Transfer 1998 Proceedings of 11th International Heat Transfer Conference, Kyongju, vol. 2, 147 (1998)
49. Varaksin, A.Y., Kurosaki, Y., Satoh, I.: An analytical investigation of turbulence reduction by small solid particles. In: International Symposium on Heat Transfer Enhancement in Power Machinery. Abstracts of Papers, Moscow, Part 1, p. 34 (1995)
50. Varaksin, A.Yu., Polezhaev, Yu.V., Polyakov, A.F.: An investigation of a "gas-solid particles" heterogeneous flow. Preprint of Institute of High Temperatures, Russian Acad. Sci. (IVTAN), Moscow, **2**, 406 (in Russian) (1997)
51. Varaksin, A.Yu., Polezhaev, Yu.V., Polyakov, A.F.: Equations of pulsation motion and pulsation heat transfer of non-stokes particles in turbulent flows. Teplofiz. Vys. Temp. **36**(1), 154 (High Temp. (Engl. transl.), **36**(1), 152) (1998)
52. Volkov, E.P., Zaichik, L.I., Pershukov, V.A.: Modelirovanie goreniya tverdogo topliva (Simulation of Solid Fuel Combustion). Nauka, Moscow (1994)
53. Wang, Q., Squires, K.D.: Large eddy simulation of particle-laden turbulent channel flow. Phys. Fluid. **8**(5), 1207 (1996)

54. Yarin, L.P., Hetsroni, G.: Turbulence intensity in dilute two-phase flows – 3. The particles–turbulence interaction in dilute two-phase flow. Int. J. Multiphase Flow **20**(1), 27 (1994)
55. Yatsenko, V.P., Aleksandrov, V.V.: Measurements of the magnus force in the range of moderate Reynolds numbers. In: Proceedings of the 9th Workshop on Two-Phase Flow Predictions, Merseburg, p. 292 (1999)
56. Yuan, Z., Michaelides, E.E.: Turbulence modulation in particulate flows: a theoretical approach. Int. J. Multiphase Flow **18**(5), 779 (1992)
57. Zaichik, L.I.: Models of turbulent transfer of momentum and heat in the dispersed phase, based on equations for the second and third moments of fluctuations of velocity and temperature of particles. Inzh. Fiz. Zh. **63**(4), 404 (1992)
58. Zaichik, L.I., Pershukov, V.A.: Problems in simulation of gas-dispersion flows with combustion or phase transitions: a review. Izv. Ross. Akad. Nauk Mekh. Zhidk. Gaza. **5**, 3 (1996)
59. Zuev, Yu.V., Lepeshinskii, I.A.: The calculation of the fluctuation parameters of the phases of a dispersed dynamically nonequilibrium two-phase flow. In: Turbulentnye dvukhfaznye techeniya (Turbulent Two-Phase Flows), Tallinn, Part 1, p. 16 (1982)

Chapter 3

1. Abezgauz, G.G. et al.: Spravochnik po veroyatnostnym raschetam (Reference Book for Probability Calculations). Voenizdat, Moscow (1966)
2. Albrecht, H.-E., Borys, M., Fuchs, W.: The cross sectional area difference method – a new technique for determination of particle concentration by laser doppler anemometry. Exp. Fluid. **16**(1), 61 (1993)
3. Cartellier, A.: LDA signals in dispersed flows: discrimination and noise. In: Laser Anemometry: Advances and Applications. Proceedings of the 2nd International Conference, Strathclyde, p. 443 (1987)
4. Dubnishchev, Yu.N., Rinkevichyus, B.S.: Metody lazernoi doplerovskoi anemometrii (Methods of Laser Doppler Anemometry). Nauka, Moscow (1982)
5. Dubovik, A.S.: Fotograficheskaya registratsiya bystroprotekayushchikh protsessov (Photographic Recording of Fast Processes). Nauka, Moscow (1975)
6. Durst, F.: Review – combined measurements of particle velocities, size distributions, and concentration. Trans. ASME J. Fluids Eng. **104**(3), 284 (1982)
7. Durst, F., Melling, A., Whitelaw, J.H.: Principles and Practice of Laser-Doppler Anemometry. Academic, New York (1976)
8. Gorbis, Z.R.: Teploobmen dispersnykh skvoznykh potokov (Heat Transfer in Disperse Through Flows). Energiya, Moscow-Leningrad (1964)
9. Greated, C.A., Durrani, T.S.: Laser Systems and Flow Measurement. Plenum, New York (1977)
10. Klimenko, A.P.: Metody i pribory dlya izmereniya kontsentratsii pyli (Methods and Instruments for Measurement of Dust Concentration). Khimiya, Moscow (1978)
11. Lee, S.L., Durst, F.: On the motion of particles in turbulent duct flows. Int. J. Multiphase Flow **8**(2), 125 (1982)
12. Leonchik, B.I., Mayakin, V.P.: Izmereniya v dispersnykh potokakh (Measurements in Disperse Flows). Energiya, Moscow (1971)

13. Modarress, D., Tan, H.: LDA signal discrimination in two-phase flows. Exp. Fluid. **1**(3), 129 (1983)

14. Modarress, D., Tan, H., Elghobashi, S.: Two-component LDA measurement in a two-phase turbulent jet. AIAA J. **22**(5), 624 (1984)

15. Qiu, H.H., Sommerfeld, M.: A reliable method for determining the measurement volume size and particle mass fluxes using phase-doppler anemometry. Exp. Fluid. **13**, 393 (1992)

16. Rinkevichyus, B.S.: Lazernaya anemometriya (Laser Anemometry). Energiya, Moscow (1978)

17. Rinkevichyus, B.S.: Lazernaya diagnostika potokov (Laser Diagnostics of Flows). Izd. MEI (Moscow Institute of Power Engineering) Moscow (1990)

18. Rinkevichyus, B.S., Yanina, G.M.: The effect of the particle size on the value of signal in an optical doppler velocimeter. Radiotekh. Elektron. **18**(7), 1353 (1973)

19. Rogers, C.B., Eaton, J.K.: The effect of small particles on fluid turbulence in a flat-plate, turbulent boundary layer in air. Phys. Fluids A. **3**(5), 928 (1991)

20. Rozenshtein, A.Z.: Measurement of the fluctuation parameters of disperse flows of the "gas–solid particles" type using a laser doppler anemometer. In: Turbulentnye dvukhfaznye techeniya (Turbulent Two-Phase Flows) Tallinn, Part 2, p. 189 (1979)

21. Rozenshtein, A.Z.: Problems in laser doppler anemometry of "gas–solid particles" flows. In: Turbulentnye dvukhfaznye techeniya (Turbulent Two-Phase Flows) Tallinn, Part 1, p. 136 (1982)

22. Rozenshtein, A.Z., Samuel', K.: The use of laser doppler velocimeter for the investigation of two-phase flows of the "gas–solid particles" type. Izv. Akad. Nauk Est. SSR- Fiz. Mat. **23**(1), 58 (1974)

23. Rozenshtein, A.Z., Samuel', K.: An optoelectronic system for diagnostics of disperse flows of the "gas–solid particles" type. In: Turbulentnye dvukhfaznye techeniya (Turbulent Two-Phase Flows) Tallinn, Part 2, p. 179 (1979)

24. Saffman, M.: Automatic calibration of LDA measurement volume size. Appl. Opt. **26**, 2592 (1987)

25. Smirnov, V.I.: Laser diagnostics of turbulence, Doctoral (Phys. -Math.) Dissertation, Moscow Institute of Power Engineering (in Russian) (1997)

26. Somerscales, E.F.C.: Laser doppler velocimeter. In: Emrich, R.J., (ed.) Methods of Experimental Physics. Academic, London, vol. 18 (Fluid Dynamics, Part A) p. 93 (1981)

27. Soo, S.L.: Fluid Dynamics of Multiphase Systems. Blaisdell, Waltham (1967)

28. Varaksin, A.Yu., Polyakov, A.F.: Measurements of the velocities of large particles using laser doppler anemometers. Meas. Tekh. **8**, 716 (1998)

29. Varaksin, A.Yu., Polyakov, A.F.: Capabilities and limitations of laser doppler anemometers in studing heterogeneous flows with solid bidisperse particles. Meas. Tekh. **9**, 888 (1999)

30. Varaksin, A.Yu., Polyakov, A.F.: Laser doppler anemometer measurement of the velocity pulsations of large particles. Meas. Tekh. **6**, 563 (1999)

31. Varaksin, A.Yu., Polyakov, A.F.: Some problems associated with experimental investigation of the structure of heterogeneous flows. Teplofiz. Vys. Temp. **38**(4), 792 (High Temp. (Engl. Transl.) **38**(4), 621) (2000)

32. Varaksin, A.Yu., Ivanov, T.F., Polyakov, A.F.: Laser doppler anemometer measurement of large-particle relative concentrations. Meas. Tekh. **8**, 852 (2001)

33. Varaksin, A.Yu., Polezhaev, Yu.V., Polyakov, A.F.: Efficiency of amplitude selection of signals in a study of geterogeneous flows using a laser-doppler anemometer. Meas. Tekh. **6**, 645 (1996)
34. van de Hulst, H.C.: Light Scattering by Small Particles. Wiley, New York (1957)

Chapter 4

1. Abramovich, G.N.: The impact made by an impurity of solid particles or droplets on the structure of a turbulent gas jet. Dokl. Akad. Nauk SSSR **190**(5), 1052 (1970)
2. Abramovich, G.N., Girshovich, T.A., Krasheninnikov, S.Yu., et al.: Teoriya turbulentnykh strui (The Theory of Turbulent Jets). Nauka, Moscow (1984)
3. Boothroyd, R.G.: Flowing Gas–Solids Suspensions. Chapman & Hall, London (1971)
4. Boothroyd, R.G., Walton, P.J.: Fully developed turbulent boundary layer flow of a fine solid-particle gaseous suspension. Ind. Eng. Chem. Fund. **12**(1), 75 (1973)
5. Derevich, I.V.: The effect of impurity of large particles on the turbulent characteristics of gas suspension in channels. Prikl. Mekh. Tekh. Fiz. **2**, 70 (1994)
6. Derevich, I.V., Yeroshenko, V.M., Zaichik, L.I.: The effect of particles on turbulent flow in channels. Izv. Akad. Nauk SSSR Mekh. Zhidk. Gaza **1**, 40 (1985)
7. Doig, I.D., Roper, G.H.: Air velocity profiles in the presence of concurrently transported particles. Ind. Eng. Chem. Fund. **6**(2), 247 (1967)
8. Fleckhaus, D., Hishida, K., Maeda, M.: Effect of laden solid particles on the turbulent flow structure of a round free jet. Exp. Fluids **5**(5), 323 (1987)
9. Gore, R.A., Crowe, C.T.: Effect of particle size on modulating turbulent intensity. Int. J. Multiphase Flow **15**(2), 279 (1989)
10. Gore, R.A., Crowe, C.T.: Effect of particle size on modulating turbulent intensity: influence of radial location. In: Turbulence Modification in Dispersed Multiphase Flows, ASME, vol. 80, p. 31 (1989)
11. Gore, R.A., Crowe, C.T.: Modulation of turbulence by a dispersed phase. Trans. ASME J. Fluids Eng. **113**(2), 304 (1991)
12. Hadrich, T.: Measurements of turbulent spectra of particle-laden flows under various conditions. In: Proceedings of 7th International Conference on Laser Anemometry: Advances and Applications, Karlsruhe, p. 1 (1997)
13. Hetsroni, G.: Particles–Turbulence Interaction. In: Third International Symposium on Liquid–Solid Flows, ASME, vol. 75, p. 1 (1988)
14. Hetsroni, G.: Particles–turbulence interaction. Int. J. Multiphase Flow **15**(5), 735 (1989)
15. Hetsroni, G., Sokolov, M.: Distribution of mass, velocity and intensity of turbulence in a two-phase turbulent jet. Trans. ASME J. Appl. Mech. **38**(2), 315 (1971)
16. Hosokawa, S., et al.: Influences of relative velocity on turbulence intensity in gas–solid two-phase flow in a vertical pipe. In: Third International Conference on Multiphase Flow, Lyon, Paper no. 279, p. 1 (1998)
17. Hutchinson, P., Hewitt, G., Dukler, A.E.: Deposition of liquid or solid dispersions from turbulent gas streams: a stochastic model. Chem. Eng. Sci. **26**(3), 419 (1971)

18. Kondrat'ev, L.V.: The structure of turbulent flow of gas suspension in the wall region of a pipe. Inzh. Fiz. Zh. **55**(6), 1029 (1988)
19. Kulick, J.D., Fessler, J.R., Eaton, J.K.: Particle response and turbulence modification in fully developed channel flow. J. Fluid Mech. **277**, 109 (1994)
20. Laats, M.K., Frishman, F.M.: The development of the methodics and investigation of turbulence intensity at the axis of two-phase turbulent jet. Fluid Dyn. USSR **8**, 153 (1973)
21. Lee, S.L., Durst, F.: On the motion of particles in turbulent duct flows. Int. J. Multiphase Flow **8**(2), 125 (1982)
22. Liljegren, L.M.: The effect of mean fluid velocity gradient on the streamwise velocity variance of a particle suspended in a turbulent flow. Int. J. Multiphase Flow **19**(3), 471 (1993)
23. Longmire, E.K., Eaton, J.K.: Structure of a particle-laden round jet. J. Fluid Mech. **236**, 217 (1992)
24. Maeda, M., Hishida, K., Furutani, T.: Optical measurements of local gas and particle velocity in an upward flowing dilute gas–solid suspension. In: Polyphase Flow and Transport Technology. Century 2 ETC, San Francisco, p. 211 (1980)
25. Maeda, M., Hishida, K., Furutani, T.: Velocity distributions of air–solid suspension in upward pipe flow (effect of particles on air velocity distribution). Trans. JSME B **46**, 2313 (1980)
26. Reddy, K.V.S., Pei, D.C.T.: Particle dynamics in solids–gas flow in a vertical pipe. Ind. Eng. Chem. Fund. **8**(3), 490 (1969)
27. Rogers, C.B., Eaton, J.K.: The behavior of small particles in a vertical turbulent boundary layer in air. Int. J. Multiphase Flow **16**(5), 819 (1990)
28. Rogers, C.B., Eaton, J.K.: The effect of small particles on fluid turbulence in a flat-plate, turbulent boundary layer in air. Phys. Fluids A **3**(5), 928 (1991)
29. Sato, Y., Hishida, K., Maeda, M.: Particle-laden turbulent flow with transverse magnetic field in a vertical channel. In: Experimental and Computational Aspects of Validation of Multiphase Flow CFD Codes, ASME, vol. 180, p. 93 (1994)
30. Shuen, J.S., Solomon, A.S., Zhang, Q.F., Faeth, G.M.: Structure of particle-laden jet: measurements and predictions. AIAA J. **23**(3), 396 (1985)
31. Simonin, O., Deutsch, E., Boivin, M.: Comparison of large eddy simulation and second-moment closure of particle fluctuation motion in two-phase turbulent shear flow. In: Proceedings of 9th Symposium on Turbulent Shear Flows, Kyoto, p. 1521 (1993)
32. Soo, S.L., Trezek, G.J.: Turbulent pipe flow of magnesia particles in air. Ind. Eng. Chem. Fund. **5**(3), 388 (1966)
33. Soo, S.L., Ihrig, H.K., El Kouh, A.F.: Experimental determination of statistical properties of two-phase turbulent motion. Trans. ASME J. Basic Eng. **82**(3), 609 (1960)
34. Soo, S.L., Trezek, G.J., Dimick, R.C., Hohnstreiter, G.F.: Concentration and mass flow distributions in a gas–solid suspension. Ind. Eng. Chem. Fund. **3**(2), 98 (1964)
35. Tsuji, Y., Morikawa, Y.: LDV measurements of an air–solid two-phase flow in a horizontal pipe. J. Fluid Mech. **120**, 385 (1982)
36. Tsuji, Y., Morikawa, Y., Shiomi, H.: LDV measurements of an air-solid two-phase flow in a vertical pipe. J. Fluid Mech. **139**, 417 (1984)

37. Tsuji, Y., Morikawa, Y., Tanaka, T., Kazimine, T., Nishida, S.: Measurement of an axisymmetric jet laden with coarse particles. Int. J. Multiphase Flow **14**, 565 (1988)
38. Varaksin, A.Yu., Polyakov, A.F.: Experimental study of fluctuations of particle velocity in a turbulent flow of air in a pipe. Teplofiz. Vys. Temp. **38**(5), 792 (2000) (High Temp. (Engl. transl.) **38**(5), 764)
39. Varaksin, A.Yu., Polyakov, A.F.: The distribution of the velocities of doubly dispersed particles in a downward turbulent flow of air in a pipe. Teplofiz. Vys. Temp. **38**(2), 343 (2000) (High Temp. (Engl. transl.) **38**(2), 343)
40. Varaksin, A.Yu., Zaichik, L.I.: The effect of a fine divided impurity on the turbulence intensity of a carrier flow in a pipe. Teplofiz. Vys. Temp. **36**(6), 1004 (1998) (High Temp. (Engl. transl.) **36**(6), 983)
41. Varaksin, A.Yu., Zaichik, L.I.: Effect of particles on carrier flow turbulence. Thermophys. Aeromech. **7**(2), 237 (2000)
42. Varaksin, A.Yu., Polezhaev, Yu.V., Polyakov, A.F.: An investigation of a "gas–solid particles" heterogeneous flow. In: Preprint of Institute of High Temperatures, Russian Academy of Science (IVTAN), Moscow, no. 2-406 (1997) (in Russian)
43. Varaksin, A.Yu., Polezhaev, Yu.V., Polyakov, A.F.: Experimental investigation of the effect of solid particles on turbulent flow of air in a pipe. Teplofiz. Vys. Temp. **36**(5), 767 (1998) (High Temp. (Engl. transl.) **36**(5), 744)
44. Varaksin, A.Yu., Polezhaev, Yu.V., Polyakov, A.F.: The effect of the concentration of particles on the intensity of pulsations of their velocities under conditions of turbulent flow of suspension of matter in gas in a pipe. Teplofiz. Vys. Temp. **37**(2), 343 (1999) (High Temp. (Engl. transl.) **37**(2), 321)
45. Wang, Q., Squires, K.D.: Large eddy simulation of particle-laden channel flow. Phys. Fluids **8**(5), 1207 (1996)
46. Wang, L.P., Stock, D.E.: Dispersion of heavy particles by turbulent motion. J. Atmos. Sci. **50**(13), 1897 (1993)
47. Yarin, L.P., Hetsroni, G.: Turbulence intensity in dilute two-phase flows – 3. The particles–turbulence interaction in dilute two-phase flow. Int. J. Multiphase Flow **20**(1), 27 (1994)
48. Yarin, L.P., Hetsroni, G.: Turbulence intensity in dilute two-phase flows – 1. Effect of particle-size distribution on the turbulence of the carrier fluid. Int. J. Multiphase Flow **20**(1), 1 (1994)
49. Yuan, Z., Michaelides, E.E.: Turbulence modulation in particulate flows: a theoretical approach. Int. J. Multiphase Flow **18**(5), 779 (1992)
50. Zaichik, L.I., Alipchenkov, V.M.: Kinetic equation for the probability density function of velocity and temperature of particles in an inhomogeneous turbulent flow: analysis of flow in a shear layer. Teplofiz. Vys. Temp. **36**(4), 596 (1998) (High Temp. (Engl. transl.) **36**(4), 572)
51. Zaichik, L.I., Pershukov, V.A.: Problems in simulation of gas-dispersion flows with combustion or phase transitions: a review. Izv. Ross. Akad. Nauk Mekh. Zhidk. Gaza **5**, 3 (1996)
52. Zaichik, L.I., Varaksin, A.Yu.: Effect of the wake behind large particles on the turbulence intensity of carrier flow. Teplofiz. Vys. Temp. **37**(4), 683 (1999) (High Temp. (Engl. transl.) **37**(4), 655)

Chapter 5

1. Balanin, B.A.: The effect of reflected particles on the carryover of mass under conditions of two-phase flow past a body. Izv. Akad. Nauk SSSR Mekh. Zhidk. Gaza 5, 193 (1984)
2. Balanin, B.A., Lashkov, V.A.: Plane wedge drag in a two-phase flow. Izv. Akad. Nauk SSSR Mekh. Zhidk. Gaza 2, 177 (1982)
3. Balanin, B.A., Zlobin, V.V.: Experimental investigation of aerodynamic drag of simple bodies in a two-phase flow. Izv. Akad. Nauk SSSR Mekh. Zhidk. Gaza 3, 159 (1979)
4. Davydov, Yu.M., Nigmatulin, R.I.: Calculation of external heterogeneous flow of gas with droplets or particles past blunt bodies. Dokl. Akad. Nauk SSSR 259(1), 57 (1981)
5. Davydov, Yu.M., Enikeev, I.Kh., Nigmatulin, R.I.: Calculation of flow of gas with particles past blunt bodies in view of the effect of reflected particles on the flow of gas suspension. Prikl. Mekh. Tekh. Fiz. 6, 67 (1990)
6. Dombrovsky, L.A.: The inertial sedimentation of particles from a gas-dispersion flow in the neighborhood of the stagnation point. Teplofiz. Vys. Temp. 24(3), 558 (1986)
7. Dombrovsky, L.A., Yukina, E.P.: The critical conditions of inertial sedimentation of particles from a gas-dispersion flow in the neighborhood of the stagnation point. Teplofiz. Vys. Temp. 21(3), 525 (1983)
8. Dombrovsky, L.A., Yukina, E.P.: The critical conditions of inertial sedimentation of particles from a gas-dispersion flow in the neighborhood of the stagnation point: the effect of injection. Teplofiz. Vys. Temp. 22(4), 728 (1984)
9. Dyban, E.P., Epik, E.Ya.: Teplomassoobmen i gidrodinamika turbulizovannykh potokov (Heat and Mass Transfer of Turbulized Flows). Naukova Dumka, Kiev (1985)
10. Kudryavtsev, N.A., Mironova, M.V., Yatsenko, V.P.: Cross two-phase flow past a cylindrical surface. Inzh. Fiz. Zh. 59(6), 917 (1990)
11. Michael, D.H., Norey, P.W.: Particle collision efficiencies for a sphere. J. Fluid Mech. 37(Pt. 3), 565 (1969)
12. Morsi, S.A., Alexander, A.J.: An investigation of particle trajectories in two-phase flow systems. J. Fluid Mech. 55(Pt. 2), 193 (1972)
13. Naumov, V.A.: Calculation of laminar boundary layer on a plate in view of lift forces acting on dispersed impurity. Izv. Akad. Nauk SSSR Mekh. Zhidk. Gaza 6, 171 (1988)
14. Nigmatulin, R.I.: Osnovy mekhaniki geterogennykh sred (The Fundamentals of Mechanics of Heterogeneous Media). Nauka, Moscow (1978)
15. Osiptsov, A.N.: The structure of laminar boundary layer of dispersed impurity on a flat plate. Izv. Akad. Nauk SSSR Mekh. Zhidk. Gaza 4, 48 (1980)
16. Osiptsov, A.N.: Investigation of zones of unrestricted growth of particle concentration in dispersed flows. Izv. Akad. Nauk SSSR Mekh. Zhidk. Gaza 3, 46 (1984)
17. Osiptsov, A.N.: The motion of dust-laden gas in the initial region of flat channel and round pipe. Izv. Akad. Nauk SSSR Mekh. Zhidk. Gaza 6, 80 (1988)
18. Rogers, C.B., Eaton, J.K.: Particle response and turbulent modification in a flat plate turbulent boundary layer. In: Turbulence Modification in Dispersed Multiphase Flows, ASME, vol. 80, p. 15 (1989)

19. Rogers, C.B., Eaton, J.K.: The behavior of small particles in a vertical turbulent boundary layer in air. Int. J. Multiphase Flow **16**(5), 819 (1990)
20. Rogers, C.B., Eaton, J.K.: The effect of small particles on fluid turbulence in a flat-plate, turbulent boundary layer in air. Phys. Fluids A **3**(5), 928 (1991)
21. Saltanov, G.A.: Sverkhzvukovye dvukhfaznye techeniya (Supersonic Two-Phase Flows). Vysheishaya Shkola, Minsk (1972)
22. Saltanov, G.A.: Neravnovesnye i nestatsionarnye protsessy v gazodinamike odnofaznykh i dvukhfaznykh sred (Nonequilibrium and Unsteady-State Processes in the Gas Dynamics of Single-Phase and Two-Phase Media). Nauka, Moscow (1979)
23. Spokoyny, F.E., Gorbis, Z.R.: Sedimentation of finely divided particles from a cooled gas flow on a heat transfer surface subjected to cross flow: characteristic features. Teplofiz. Vys. Temp. **19**(1), 182 (1981)
24. Tsirkunov, Yu.M.: The effect of a viscous boundary layer on the sedimentation of particles under conditions of gas suspension flow past a sphere. Izv. Akad. Nauk SSSR Mekh. Zhidk. Gaza **1**, 59 (1982)
25. Tsirkunov, Yu.M.: Gas-Particle Flows around Bodies – Key Problems, Modeling and Numerical Analysis. In: Proceedings of the 4th International Conference on Multiphase Flow (ICMF'01), New Orleans, P607, p. 1 (2001) (CD-ROM)
26. Varaksin, A.Yu.: Modification of turbulence of flow by solid particles. In: Nauchnye osnovy tekhnologii XXI veka (Scientific Foundations of Technologies of the XXI Century). UNPTs Energomash, Moscow, p. 98 (2000)
27. Varaksin, A.Yu.: Investigation of the structure of "gas–solid particles" turbulent heterogeneous flows. In: Doctoral (Phys.–Math.) Dissertation, Joint Institute of High Temperatures, Russian Academy of Science, Moscow (2001) (in Russian)
28. Varaksin, A.Yu., Ivanov, T.F.: Effect of the particles concentration on their velocity distributions for heterogeneous flow near blunted body. In: Proceedings of the 4th International Conference on Multiphase Flow (ICMF'01), New Orleans, P793, p. 1 (2001) (CD-ROM)
29. Varaksin, A.Yu., Mikhatulin, D.S., Polezhaev, Yu.V., Polyakov, A.F.: Measurements of velocity fields of gas and solid particles in the boundary layer of turbulized heterogeneous flow. Teplofiz. Vys. Temp. **33**(6), 915 (1995) (High Temp. (Engl. transl.), **33**(6), 911)
30. Varaksin, A.Yu., Mikhatulin, D.S., Polezhaev, Yu.V., Polyakov, A.F.: Measurements of the velocity distributions in an air–solid two-phase boundary layer. In: Proceedings of 10th Symposium on Turbulent Shear Flows, Pennsylvania, vol. 3, p. 79 (1995)
31. Varaksin, A.Yu., Polezhaev, Yu.V., Polyakov, A.F.: Velocity and its fluctuations distributions at the gas–solid turbulent boundary layer. In: Proceedings of 4th World Conference on Experimental Heat Transfer, Fluid Mechanics, and Thermodynamics, Brussels, vol. 3, p. 1417 (1997)
32. Vittal, B.V.R., Tabakoff, W.: AIAA J. **5**, 648 (1987)

Index

Springer Series on
ATOMIC, OPTICAL, AND PLASMA PHYSICS

Springer Series on
ATOMIC, OPTICAL, AND PLASMA PHYSICS